Heritage and Wellbeing

Heritage and Wellbeing

The Impact of Heritage Places on Visitors' Wellbeing

Faye Sayer

OXFORD
UNIVERSITY PRESS

OXFORD
UNIVERSITY PRESS

Great Clarendon Street, Oxford, OX2 6DP,
United Kingdom

Oxford University Press is a department of the University of Oxford.
It furthers the University's objective of excellence in research, scholarship,
and education by publishing worldwide. Oxford is a registered trade mark of
Oxford University Press in the UK and in certain other countries

Published in the United States of America by Oxford University Press
198 Madison Avenue, New York, NY 10016, United States of America

British Library Cataloguing in Publication Data
Data available

Library of Congress Control Number: 2024930875

ISBN 9780198871644

DOI: 10.1093/9780191914539.001.0001

Printed and bound in the UK by
Clays Ltd, Elcograf S.p.A.

Acknowledgements

This book and the research it describes are dear to my heart; archaeology and history have provided me, a person with dyslexia, a happy place, a place to make friends, to grow in confidence, to do something I love, and to escape difficulties. Doing this research, my job, and collaborating with communities and colleagues has given me joy and enabled me to flourish. Thank you to all friends and colleagues along the way. Reflecting on writing this book has been both good and bad for my own wellbeing, for both personal and professional reasons.

I cannot express my gratitude enough to my in-house editor, my father-in-law Phil Sayer. Thank you for being a meticulous editor, for your critical eye for detail and fact checking. He is still talking to me though; I will stop saying sorry one day.

A special thank you to Mark Pickel, a special appreciation for your friendship, continued support, guidance, and for being inspiring; for being a critical friend whom I trust and value as a colleague and friend. I am so grateful for your first email and reading through my drafts. Thank you for saying the right thing at right time; I will endeavour to be a Rhino.

To all the volunteers, heritage staff, and interns who helped collect data; specific thanks to Amy Luck who has been the amazing research assistant for this project, for keeping on top of administration, brilliant organizational skills, and for being a wonderful person. To Sakshi Kanchan, the projects research assistant in India for supporting student interns. At heritage places thank you to Jean Hershner and Annie Thomas-Bubel (Nemours Estate), Mark Pickel and Ellen Rickman (Biltmore Estate), Sally Ambrose, Thomas Hendry, Jonathan Fish, and Hatty O'Sullivan (Chatsworth Estate and Devonshire Group), Michele Collings (Priory Gardens and Dunstable Town Council), Suzanne Bardgett and Mike Peirson (Imperial War Museum, Duxford), Jane Finley and Kelly Robinson (Dulwich Picture Gallery), and to Archaeological Survey India for their support and understanding permission. To various interns including Evelyn Gilks, Kelsey Marriott, Tom Blackburn (Manchester Metropolitan University, UK), Bodala Durga Pravallika, Janani T, Medidi Maidhili, Sri Satya Ruchita Mandavilli, Venumbaka Bhuvidha, Barkha Rohra,

Aishwarya Karkal, Yashashwini L (JD School of Design, India), and at National Museum Institute, Kajal Sharma. To all the people I have had opportunity to collaborate with for truly caring about peoples' wellbeing especially Anna Bergqvist, Shawn Sobers, Leigh Chambers.

Thank you to the British Academy for a Small Grant (SRG1819\190822), derived from the Academy's partnership with the Department for Business, Energy, and Industrial Strategy, the Chittick Fund and the Albert Reckitt Fund that supported this research.

To my wonderful husband Rick, my best friend and the person I trust, who makes me happy and challenges me. Thank you for holding the fort while I have disappeared to collect data and to try to figure out what it all means. Thank you for your support as a psychologist in residence for introducing me to behaviour change models. I look forward to working together on our next project. My love and thanks to all the family and to our beautiful, kind, and manic children Hector, Temperance, and Honor, keep being busy and challenging, be as proud of yourselves as we are of you.

Contents

List of Figures

List of Acronyms

BE	Biltmore Estate
BCW	Behaviour Change Wheel
BF	Bangalore Fort
CE	Chatsworth Estate
CGMAG	Chandigarh Government Museum and Art Gallery
COM-B/	Capabilities, Opportunities and Motivations Behaviour Change
DCMS	Department for Digital, Culture, Media, and Sport
DPG	Dulwich Picture Gallery
EU	European Union
GAD	Generalized Anxiety Disorder questionnaire
GDPR	General Data Protection Regulation
GNP	Gross National Product
HE	Historic England
IWM	Imperial War Museum Duxford
MWS	Modified Wellbeing Scale
NAW	National Accounts for Wellbeing
NEF	New Economics Foundation
NLHF	National Lottery Heritage Fund
ONS	Office for National Statistics
PANAS	Positive and Negative Affect Schedule
PG	Priory Gardens
PHQ	Patient Health Questionnaire
PIS	Participant Information Sheet
RA	Research Associate
SD	Standard Deviation
SIAW	Self-Identified Aims Wheel
SROI	Social Return of Investment
TSSP	Tipu Sultan's Summer Palace
UN	United Nations
UNESCO	United Nations Educational, Scientific, and Cultural Organization
VAG	Venkatappa Art Gallery
VAS	Visual Analogy Scale
WEMWBS	Warwick–Edinburgh Mental Wellbeing Scale
WHO	World Health Organization

1
Introduction

Heritage and Wellbeing examines the role heritage plays in creating healthier societies. It considers how heritage can support political agendas that positively impact on individual and community wellbeing. The book seeks to explore, through a range of international case studies, how heritage can improve societal wellbeing and how people's experiences of and participation in heritage affect their wellbeing and mental health. The research presented will provide evidence of how public engagement in and with 'cultural' heritage can provide a mechanism to support healthier and happier societies. It concludes that heritage should be regarded by professionals, organizations, and policy makers as a key player in policy and practice related to improving wellbeing and mental health.

Research into wellbeing, specifically heritage wellbeing, is an emerging area of study, and there is a significant gap in the understanding of how diverse heritage activities including museums, heritage sites, and community archaeological excavations can impact visitors' wellbeing. Furthermore, research has still to consider how heritage activities in a range of cultural contexts could shape individual and communal wellbeing. 'Heritage and Wellbeing' strives to rectify this disparity between wellbeing research and heritage research and determine the role that heritage can play in creating healthier societies. It offers original evidence pertaining to how different forms of public engagement with heritage activities can contribute to healthier and happier societies. Key case studies from North America, Asia, and Europe, including ten heritage places from the developed and developing world, provide a comparative assessment of the impact of diverse heritage assets on public wellbeing. The breadth of case studies enables the impact of heritage on the public's wellbeing to be contextualized, in relation to specific social, cultural, economic, and political influence factors. This book presents a significant body of new research in the emerging field of wellbeing studies and heritage wellbeing studies, offering the reader a comprehensive assessment of how different types of heritage within diverse cultural contexts can impact on mental health and wellbeing.

The chapters in 'Heritage and Wellbeing' firmly place heritage in the wider nexus of wellbeing studies. The book starts with a brief synopsis of wellbeing theory and policy, including heritage wellbeing (Chapter 1), and is followed by a discussion of the diverse mechanisms for the evaluation of heritage wellbeing (Chapter 2). The methodological chapter (Chapter 3) details the research applications of a 'double helix' evaluation strategy. Chapters 4 and 5 present evidence regarding how specific heritage places impact individuals' subjective wellbeing and provides a brief oversight of the different capabilities and indicators associated with varying heritage places. Chapter 6 frames these heritage places within diverse cultural contexts, and discusses how national cultural, political, and social arenas of the key case studies impact on both individual subjective wellbeing and the impact of demographics on wellbeing change. The conclusion in Chapter 7 provides evidence of how public engagement with heritage places, including museums, art galleries, heritage sites, and archaeological sites, impacts on wellbeing indicators and considers key themes and patterns that will inform both future heritage wellbeing research and health and wellbeing policy.

The research described in 'Heritage and Wellbeing' suggests that by applying a humanistic approach to heritage practice and applying public health perspectives to heritage project design and evaluation, cultural heritage resources can be a proactive tool to support individual wellbeing, promote societal change and create healthier societies. It shows how engagement in heritage can positively impact individuals' subjective wellbeing and offers ideas for future strategies that heritage organizations might wish to implement within their management of heritage places to improve visitor wellbeing and support their role in creating healthier societies.

'Heritage and Wellbeing' encourages heritage professionals to adapt their approaches to public engagement to support positive mental health and wellbeing, and work together as a profession and with academia to demonstrate the value of heritage, archaeology, and history to the public, and the value of heritage for improving personal mental health and wellbeing. 'Heritage and Wellbeing' securely positions heritage as an important resource to be incorporated within national and international health and social policy frameworks.

Background

Over the last two decades the breadth of high-quality research into heritage's impact on society has flourished; more recently research into wellbeing and heritage has become an emerging field. Yet heritage wellbeing research has often failed to provide sufficient high-quality data to impact on wider health and social policy. This in part is due to several issues including: a reliance on singular case studies, focus on western contexts, lack of rigorous data collection, no standardization of data collection, no contextualization of data, and lack of application to wider policies. Significant gaps in data relating to heritage and wellbeing also exist, including investigating the unique impact diverse heritage assets have on wellbeing and the how heritage resource context impacts on wellbeing.

'Heritage and Wellbeing' came about because of the personal and professional experiences of the author, from first-hand experience and anecdotal evidence of the impact of heritage places and projects on the wellbeing of participants, visitors, and staff, and research into qualitative and quantitative evaluation of 'community' heritage projects' impact on wellbeing. Consequently, this book is based on experience and the hypothesis that heritage can make a difference to wellbeing, by stimulating positive changes to mental health and wellbeing within diverse societies. However, for this hypothesis to be proved, heritage needs to demonstrate its worth and be on a level playing field with other disciplines that have already established their social benefits such as geography, art, and literature. The only way to prove heritage's worth to wellbeing and subsequently position heritage as a 'wellbeing'-generating equal within the humanities sector is to provide rigorous and robust evidence of its impact. This evidence must be generated using pre-accepted public health methods for measuring mental health and wellbeing, applied to an international range of diverse heritage projects. Consequently, evaluation of heritage wellbeing requires the application of a mixed method approach that utilizes qualitative and quantitative methods, including Positive and Negative Affect Schedule and Visual Analogue Scale, to provide evidence-based conclusions (Abdallah et al. 2011).

The foundation of the research described in this book is the development of an accurate data capture strategy. This strategy was piloted by the author on five projects in the United Kingdom and one in the United States and applied to a diverse range of publicly accessible heritage projects and resources (Sayer 2015, 2018). The pilot study has enabled the development of a globally applicable evaluative method of data gathering which can be successfully applied to a wide range of heritage projects. This method has facilitated the preliminary identification of specific elements within heritage practice that produced the greatest positive changes to wellbeing, for example: active physical engagement, collaboration, co-creation, community leadership, ownership, personal reward, discovery, learning, and environmental and social contexts (Sayer 2015, 2018). The method applied to this research will enable the heritage profession to understand how public engagement in heritage can impact wellbeing and mental health. This will enable the heritage profession to demonstrate the distinctive methodological practices that qualify heritage to be a significant player in positively impacting wellbeing, happiness, and mental health.

The Theoretical Basis of Wellbeing

It is impossible to discuss heritage's impact on wellbeing without first engaging in the underpinning epistemological frameworks and theoretical paradigms from which wellbeing, wellbeing hypotheses, and wellbeing research methods stem. Paramount to this discussion is understanding the concept of wellbeing and the historiography and phenomenology behind its definition. This includes discussing the theoretical principles underpinning current wellbeing measurements and models of assessment, and critically examining issues with current approaches to wellbeing research. More detail regarding measuring wellbeing is provided in Chapter 2.

For both Plato and Aristotle, the central problem of ethics was the achievement of happiness. By 'happiness' (the usual English translation of the Greek term *eudaimonia*), they did not mean a pleasant state of mind but rather 'happiness that is observable,' that which is constructed, produced, and circulated—similar to our modern concept of wellbeing (Knight, La Placa, and McNaught 2014: 7, Tovar-Restrepo 2012, Evans 2018, Bache and Scott 2018, Scott 2012, Rose 2009). In essence, wellbeing is an ever-changing social construct and a judgement of one's life, that can be shaped by cultural context and social norms (Mehmet 2011, Knight et al. 2014: 7, Scott 2012: 6).

Subjective wellbeing has been divided into hedonic and eudaimonic wellbeing, with both linked to the concept of happiness (Ryan et al. 2001,

Scott 2012). Hedonic wellbeing is linked to idea of maximizing pleasure and absence of pain, but also cultivation of knowledge and freedom to pursue activities (Bache and Scott 2018: 11, Bradley 2015). Hedonic wellbeing considers happiness as linked to the presence of positive emotions and absence of negative emotions (Lucas 2018). Eudaimonic wellbeing has been associated with the fulfilment of certain virtues and goals that leads to a good life and enables people to flourish (Scott 2012, Bache and Scott 2018, Evans 2018). Some theoretical classical schools of thought virtues linked to eudaimonic wellbeing have included cultivation of knowledge and skills to pursue pleasure and avoid pain (Epicurean), resilience (Stoics), and development of character and intellectual virtues (Aristotelian) (Annas 1998, Rabbas et al. 2015, Bache and Scott 2018: 9). Eudaimonic wellbeing is linked to satisfaction, having a good life and quality of life, spiritual wellbeing, happiness, and the possession of skills, wisdom, and excellence that will enable people to flourish (Bache and Scott 2018: 9). It is also linked to the attainment of certain virtues that enable people to flourish, including courage, justice, moderation, greatness of soul, hospitality, knowledge and perception, proper judgement, and practical wisdom (Nussbaum and Sen 1993: 245–246, Bache and Scott 2018: 9, Scott 2012, Sen 1980, 1993).

In recent decades wellbeing has been associated with physical and mental health, for example: 'health is a state of complex physical, mental and social wellbeing, and not merely the absence of disease or infirmary' (WHO 1984). Consequently, individual and societal wellbeing and good health are perceived to be affected by external conditions, and these external conditions impact on how individuals function in society (Abdallah et al. 2017). Wellbeing has been associated with quality of life, prosperity, and sustainable thriving communities, and to liberal concepts such as freedom, liberty, and rights (Abdallah et al. 2017, Reily et al. 2018: 5, Scott 2012: 38). The factors considered to impact wellbeing have included: income, work, education, mental and physical health, social relationships, loneliness, social isolation, the environment, and freedom (Bache and Scott 2018). It has been suggested that positive wellbeing requires equality in supporting and creating conditions in which these attributes can thrive and human capabilities can flourish (Abdallah et al. 2017, Bache and Scott 2012: 39). Positive wellbeing has been associated with feelings of contentment, enjoyment, self-confidence, good relationships, engagement in the world, resilience, sense of ownership and people's enthusiasm do things, and has been linked to high levels of personal happiness (Reily et al. 2018, McNaught and Knight 2011, Seligman 2011, Abdallah et al. 2017,

Scott 2012). Wellbeing is often used interchangeably with happiness, which is associated with good mental health, personal satisfaction, and positive emotional states (Griffin 1988, Seligman 2011, McNaught and Knight 2011, Seligman 2011). As a consequence of this, a person could be deemed to have positive wellbeing if they have positive perceptions of themselves, their social relationships, and the surrounding environment (Lyubomirsky et al. 2005, Nussbaum 2005, Jordon 2008).

Previous research into wellbeing largely sought to measure wellbeing through 'quantifiable variables' and often objective indicators of wellbeing (Scott 2012). Objective indicators included material and social circumstances in which individual or community wellbeing is measured and is often linked to Gross National Product indicators such as economy, income, housing (Scott 2012). This objective approach does not include an evaluation of hedonic wellbeing, which is based on individual subjective perception of emotions, for example anxiety. As a result, social scientists and psychologists asserted that was not possible to measure wellbeing and happiness through an objective approach, rather they championed a subjective and social indicator approach to measuring wellbeing grounded in quality of life (Rowles 1971, Scott 2012). Hence the measurement of eudaimonic wellbeing has rejected a solely material approach, based on possession of resources, in favour of a capabilities approach (Scott 2012). The capabilities approach is grounded in a virtues-based view to humans flourishing, in which wellbeing is subjectively measured based on what people are and feel able to do (Scott 2012: 22, Sen 1980, 1993, Nussbaum 2005, Nussbaum and Sen 1993). This approach has been linked to the theory of needs, which asserts that people have psychological needs, which are both primary and immediate including safety, belonging and esteem, autonomy, relationships, education, security, and physical health (Maslow 1943, Doyal and Gough 1991). Measuring eudaimonic wellbeing through a capabilities approach could be likened to a measurement of social and personal wellbeing components and indicators, for example the National Accounts of Wellbeing Indicators or New Economic Foundation Five Ways to Wellbeing (Ryff 1989, Aked et al. 2008, Michaelson et al. 2009).

The quantifiable wellbeing measurements trend correlated, in part, with the development of the definition of wellbeing, which was and perhaps still is dominated by a positivist epistemological angle (Knight et al. 2014, Scott 2012). This theoretical approach to wellbeing resulted in economists, psychologists, health scientists, and social scientists devising models to quantify wellbeing (Abdallah et al. 2011). These models defined and categorized wellbeing, which was used interchangeably with four specific positivist and quantifiably measurable characteristics, which included (Knight et al. 2014: 8–12):

1. *Quality of Life*: The physical rather than psychological external observations and evaluations by people regarding how satisfied they are with their external conditions (Mathews and Izquierdo 2009).
2. *Life Satisfaction*: Measurement is influenced by psychological discourse relating to feeling and emotions and is based on how people are feeling about their lives in various dimensions, including time (past, present, future); interpersonal relations (social and specific) and self (self-esteem, self-perceptions, self-concept) (Knight et al. 2014: 11, Graham 2011).
3. *Subjective Wellbeing*: Subjective wellbeing based on the notion of happiness made scientific (Knight et al. 2014: 11). Measurement involves people's self-reported personal wellbeing, including the personal belief that something is desirable, pleasant, and good (Graham 2011, Diener 2009a, 2009b, Diener et al. 1995). Measurement is acquired through different and open-ended elements linked to satisfaction including: health, work, education, life itself.
4. *Happiness* Happiness is based on ideas of contentment and pleasure through experience (Graham 2011). Measurement is based on observing and collecting data based on economic characteristics of happiness and can include but are not limited to income metrics (Graham 2011).

The measurement of subjective wellbeing is regularly based on the personal assessment of an individual's quality of life, which can include: how we are doing; how we experience life; our quality of life; our good physical and mental and emotional health; and our role as part of a community (Abdallah et al. 2017, Scott 2012: 28). Wellbeing assessments are frequently centred around the four characteristics of wellbeing (subjective wellbeing, happiness, life satisfaction, and quality of life), and subsequently indexes, models, and measurements are based on these definitions and on positivist assumptions regarding wellbeing. These assumptions include: valuations of wellbeing are fact and as such neutral and objective; wellbeing is measurable, predictable, and controllable, and understood in terms of cause-and-effect models; wellbeing produces hard, measurable data; wellbeing is orderly, systematized, and knowable through its four category indicators; wellbeing can be observed at both individual and societal levels (Scott 2012). Resultantly, the majority of wellbeing definitions and measurements assume that humans are rational, and subjective wellbeing can be standardized (Bache and Scott 2018, Scott 2012, Austin 2018).

Recent studies into wellbeing have frequently been framed around subjective wellbeing: the self-reported individual emotional reactions and judgements related to life satisfaction and happiness (Halpern 2015, Christie and Nash 1998, Kahneman et al. 1999). Consequently, wellbeing can be regarded as a subjective experience of happiness and positive emotions (Halpern 2015, Scott 2012, Bradley 2015). This concept of wellbeing has provided alternative psychological and social science methodologies that diverge from the previous positivist concepts of wellbeing. These alternative qualifiable approaches to defining and measuring wellbeing regard it as both subjective and objective, contextually independent and dependent, and in a constant state of flux (Putman 2006). These distinctive approaches to definitions and methods of assessment indicate that wellbeing requires in-depth unpacking in order to explore the limitations of existing research and methodological frameworks. To accomplish a critical approach to wellbeing, the diversity of wellbeing needs to be considered at individual, communal, local, and societal levels. The unpacking of wellbeing requires the introduction of phenomenological, social constructivist and post-structuralist theories to 'wellbeing' discussions (Knight et al. 2014). This enables wellbeing to adopt a humanistic epistemological angle, in which understanding and measuring wellbeing can deviate from 'economically -centric' notions of income and growth, and mathematical equations (Sedlacek 2011, Scott 2012).

The development of alternative ways of measuring wellbeing required researchers to view wellbeing through an anthropological and ethnographical lens, and to understand that feelings and emotions are also linked to hedonic and eudaimonic wellbeing. Consequently, understanding wellbeing requires understanding moral codes, context, and relationships, and the dynamic relationship of wellbeing to cultural activities including philosophy, art, religion, literature, and communal narratives, which cannot be measured by standardized quantitative methods alone (Thin 2009, Bok 2010, Scott 2012, Evans 2018, Bache and Scott 2018: 15). The unpacking of the complex and dynamic nature of individual wellbeing is at the heart of 'Heritage and Wellbeing', and its critical viewing through scientific and humanistic lenses of the impact heritage resources have within specific cultural contexts.

Wellbeing is acutely difficult to accurately define, due in part to its intangible and subjective nature (Griffin 1988). For the purposes of this research, wellbeing is defined as a facet of an individual's overall self-perception, attributed to a combination of personal, social, and psychological variables, values and capabilities (Dodge et al. 2012, Scott 2012). It will be seen through the lens of both positivist and negativist epistemological angles, with notions of wellbeing being socially contextualized through a combination of quantitative and qualitative measurements. Wellbeing will be measured through quantitative and qualitative assessment based on the subjective rather than objective (economy, housing, income) characteristics and capabilities of hedonic and eudaimonic wellbeing, including happiness, life satisfaction and subjective personal, social, and psychological wellbeing. This holistic approach is based on the hypothesis that wellbeing is in a constant state of flux, but specific actions including types of engagement at heritage sites and cultural contexts influence an individual's state of wellbeing.

This research, by understanding the multifaceted phenomenon of wellbeing through capabilities and social values measurement of wellbeing, will help educate and inform the public, professionals, and the politicians about what wellbeing is in relation to heritage settings, and how these settings affect subjective hedonic and eudaimonic wellbeing (Froggett 2011). The research seeks to use more sophisticated ways of thinking and measuring wellbeing in the distinct setting of heritage places and provide hard evidence to inform debate.

Wellbeing Policy and Practice

Wellbeing was first articulated in policy in the mid-twentieth century (Knight et al. 2014: 19, Scott 2012). Following the Second World War, two specific strands of wellbeing policy developed. The first was a western-centric objective and material-based approach centred on the prevailing dominance of the modern capitalist paradigm, which consequently linked wellbeing to neoliberal economic policies (Scott 2012: 3). In this context, wellbeing was politically articulated in the context of economic growth, income, prosperity, welfare, and gross national product (Knight et al. 2014: 19, Scott 2012: 17). In the United Kingdom and Europe, wellbeing was objectively and quantifiably linked to economic development and living standards.

The second was a social approach to wellbeing and policy based on the subjective non-western 'holistic' and social 'anthropological' concepts of wellbeing, in which wellbeing was linked to social capital, related to sustainable, social, and non-economically focused development of an individual's quality of life (Mathews and Izquierdo 2009, Ura et al. 2012: 7, Bache and Scott 2018: 30). For example, in Bhutan wellbeing was linked to happiness and was regarded as more important to sustainable development than the economy, and as a result was linked to wider social and healthcare policies (Ura et al. 2012: 8, Scott 2012: 3). Social values associated with wellbeing have been regarded as critical to the continued development and growth of countries around the world, perceived as essential goals to sustainable global development agendas, and linked to economic development, resource management, and equality in the provision of education and health care. Supporting and improving positive social values has been associated with reducing wellbeing inequality and creating sustainable positive wellbeing improvements.

Social values are associated with a range of intangible principles, actions, feelings, and reactions that affect people's lives and social dynamics, and can be defined as 'the quantification of the relative importance that people place on the changes they experience in their lives' (Social Value UK 2018). Social values attribute importance to feelings and actions, and these values play a significant role in society, both to individuals and to communities. These values include 'softer attributes' characteristics that impact on individuals. These traits are often intangible (qualitative) and can include a 'sense of self', 'sense of belonging', 'connection to community', 'building confidence', 'feeling happy', and 'being relaxed.' Social values influence actions, reactions, equality, democracy, and morals, and have huge potential to change the way we interact and understand ourselves and the environment we live in. Social values have been linked to maintaining social order and are associated with the way a person experiences and appreciates their lives in the present. Social value has been perceived as having a quantifiable economic value, regarded as having 'additional benefit above direct purchasing of goods, services and outcomes' and has been considered in terms of what benefit it offers to the community based on cost benefit models (Dayson 2013).

The improvement of wellbeing has been linked to social values, and supporting wider economic, social, and human capital, that can impact a range of national and global issues, including individual standards of living, education, health, economics, and stability of a country. The broad range of social value impacts explains the rationale behind why international organizations, and national and local governments are interested and invest in research into societal values of populations and develop specific programmes to positively impact on these values. National and international government and non-government agencies strive to track changes in these social values, so as to better understand them and to deliver policies, programmes, and actions that positively alter or develop society. In the United Kingdom, the development of the Public Services (Social Value Act) 2012 was based on national social value research. This required people who commission public services to consider how they could secure wider social benefits (Public Services 2012). The Social Value Act was a tool to improve value for money by designing better 'value' social and economic services for the community. The social value awards for individuals and organizations were introduced to celebrate good practice in commissioning and providing social value. For example, the Association of Greater Manchester Authorities Procurement Hub received a leadership award for its work on developing a social value policy (Cabinet Office 2013). This programme of work identified three barriers that existed for social value; this included: (i) awareness and take up, (ii) practical understanding, and (iii) measurement (Cabinet Office 2013). All salient points to remember when considering the application of social value within heritage or even wellbeing and heritage.

Since the late twentieth and early twenty-first century wellbeing has become a central theme in political discussions (Knight et al. 2014, Scott 2012). As concepts, meanings, and impacts of wellbeing were examined globally, health and wellbeing was increasingly placed at the heart of many public policy documents (Knight et al. 2014, Scott 2012). Integrating wellbeing elements became a predominant political practice in western world policy development and strategy. Notions of wellbeing have filtered through into specific and non-specific health and wellbeing policies, which have included: public health, education, economics, regeneration, and social reform. Wellbeing, lowering wellbeing inequality, and reducing wellness poverty underpins strategy in numerous international and national public policies. Wellbeing as a political practice has influenced health outcomes, public expenditure, preventative care measures, mental health strategies, and even electoral fates (Ward 2015). Consequently, wellbeing has become a political priority for many international and national organizations and governments, including both developed world countries such as the United States, United Kingdom, Australia, and developing world countries such as Bhutan and Nigeria. The United Nations and World Health Organization have investigated wellbeing at a global scale and the UK government's Office of National Statistics have attempted to capture, measure, and note changes in society's wellbeing (United Nations 2011, World Health Organization 2008, Office for National Statistics 2013). In a wide range of international and national contexts, wellbeing has become a hallmark of a 'healthy' society's development, including the societal impacts that influence wellbeing (Halton NHS Clinical Commissioning Group 2017).

Current national, international, and global government agendas for investigating wellbeing and Gross National Happiness have spurred academic and commercial organizations to consider whether their work has the potential to influence people's happiness and sense of wellbeing. This section provides an overview of the international and national policies relating to wellbeing, and where this has permeated into specific heritage sector wellbeing policies or strategies.

International and Global Policies

International organizations including the United Nations, the World Health Organization, the World Bank, the European Union, and UNESCO have engaged in wellbeing research and integrated it into their international resolutions, strategies, guidelines, and policies (United Nations 2011, UNESCO 2016, WHO 2008, European Commission 2008). Research into international health-related and national interest wellbeing has filtered into policies and strategies of international intergovernmental organizations, including the United Nations and European Union (WHO 1946, Ura et al. 2012: 13, European Union 2022). In 1948, the World Health Organization was the first international organization to formally discuss wellbeing and link it to health (WHO 1946). It suggested that health was not just the absence of biomedically defined disease but also included wellbeing (WHO 1946). It was this statement that framed discussions within international and national organizations regarding how to measure wellbeing, specifically using quantitative 'biomedical' and health care methods.

In 2011, the United Nations, an organization whose primary aim is to maintain international peace and security, adopted a General Assembly resolution (65/309) based in part of Bhutan's Gross National Happiness (GNH) approach (UN 2011). Resolution 65/309 recognized that the conditions in which people live and their lifestyles contribute to health and called for a 'holistic approach to development' that focused on promoting sustainable happiness and wellbeing (UN 2011). Subsequently, in 2012, the UN hosted an international meeting 'Happiness and Wellbeing: Defining a New Economic Paradigm', bringing together world leaders and experts to develop a new economic model based on sustainability and wellbeing.

The Organization for Economic Co-operation and Development, an international organization working with national governments, policy makers, and citizens, developed the globally applicable 'Better Life Index' in 2011 (OECD 2017). The index identified eleven essential topics for wellbeing in areas such as material living conditions and quality of life, which included: housing, income, jobs, community, education, environment, governance, health, life satisfaction, safety, and work-life balance (OECD 2017). The survey aimed to support countries to create their own Better Life Indices, and to work together to generate comparative data on wellbeing that could aid a global understanding of wellbeing. The comparative data published in the OECD 'How's Life' survey investigated the impact of these eleven topics on wellbeing and ranked countries' wellbeing, for example in 2017 Norway and Australia ranked the highest in wellbeing (OECD 2017). The research aimed to determine overarching patterns in wellbeing topics, which could be integrated into national and international policies to positively impact wellbeing.

In 2015, the United Nations launched the 2030 Agenda for Sustainable Development. This was adopted by all United Nations Member States and aimed to support peace and prosperity for people and the planet, now and into the future (UN 2015). The agenda was linked to seventeen Sustainable Development Goals (SDGs), each linked to specific strategies to improve health and education, reduce inequality, support economic growth, tackle climate change, and work to preserve oceans and forests. SDG3 'Good Health and Wellbeing' directly outlined the importance of wellbeing to agendas for peace and prosperity (UN 2015). The United Nations Educational, Scientific, and Cultural Organization (UNESCO) has incorporated wellbeing into wider global strategies. For example, in 2016 UNESCO launched its strategy on 'Education for Health and Wellbeing,' which outlined wellbeing's role in the UN's global Sustainable Development Goals. Sustainable Development Goal 3 'Good Health and Wellbeing' focused on improving health and wellbeing through youth and adolescence education and awareness programmes (UN 2015, UNESCO 2016). UNESCO formally outlined the rights of people, in every country, to access health, wellbeing, and education and indicated this goal would be achieved through partnerships and investment (UNESCO 2016). International support for achieving these Sustainable Development Goals came in the form of the Global Challenges Research Fund, which provided grants to contextually and geographically specific projects and schemes which sought to measure and improve wellbeing for communities where wellbeing and healthcare inequality was most evident (UNESCO 2016, UN 2015).

In 2021, the WHO ratified the Geneva Charter for Well-being, which outlined the foundations to support wellbeing through policy and transformative approaches for current and future generations to thrive, which included: a positive vision of health that integrates physical, mental, spiritual and social well-being; the principles of human rights, social and environmental justice, solidarity, gender and inter-generational equity, and peace; a commitment to sustainable low-carbon development grounded in reciprocity and respect among humans and making peace with nature; new indicators of success, beyond gross domestic product, that take account of human and planetary wellbeing and lead to new priorities for public spending; and a focus of health promotion on empowerment, inclusivity, equity, and meaningful participation (World Health Organization 2022). It outlined five key action areas to support wellbeing which were linked to UN SDG3 (UN 2015). These included: to design an equitable economy that serves human development within planetary boundaries; to create public policy for the common good; to achieve universal health coverage; to address the digital transformation to counteract harm and disempowerment and to strengthen the benefits; and to value and preserve the planet (World Health Organization 2022).

As a result of this global and inter-organizational focus, wellbeing is now connected to various international policies including: health, security, sustainability, reduction of poverty, and education. The infiltration of wellbeing into diverse government policies has been guided by a redefinition of health and the recognition of the requirement to adopt a holistic approach to healthcare, education, and sustainable development. Wellbeing's perceived impact on a range of development agendas has enabled it to bridge interdisciplinary boundaries and to be incorporated into broader global and national political strategies.

National Policies

Improving national wellbeing has become entwined with national goals to improve physical and mental health, and has become associated with increasing jobs, improving education, reducing healthcare costs, and increasing economic growth (Allin and Hand 2014). Political support for and policies related to wellbeing and its associated social values can impact on public perceptions of positive societal change and perform a critical role in garnering the public's support for governments and their agendas. As result, some national governments have concluded that having a positive impact on wellbeing could even affect their success in future elections (Ward 2015).

The integration of wellbeing into national policies began after the 1950s. Most of these policies sought to link wellbeing to economic development and Gross Domestic Product (Knight et al. 2014, Scott 2012). In 2009, the French Government's Commission on the Measurement of Economic Performance and Social Progress suggested alternative measures to GDP should incorporate elements of health, including indicators of individual and societal wellbeing (Stiglitz, Sen, and Fitoussi 2009).

Bhutan was one of the first countries to recognize the importance of wellbeing and suggested it should be measured through the inclusion of non-economic social aspects (Ura et al. 2012). Bhutan created the 'Gross National Happiness (GNH) Index' and undertook the first GNH survey in 2007, which was adopted in 2010 (Ura et al. 2012: 13). The GNH index used subjective and objective wellbeing measures, which included nine domains and thirty-three indicators: psychological wellbeing, health, education, time use, cultural diversity and resilience, good governance, community vitality, ecological diversity and resilience, and living standards (Ura et al. 2012: 20). The results of the GNH survey were integrated into Bhutan's national sustainable development policies and influenced economic and social policies, which focused on what happy people enjoyed and how this could be provided for everyone (Ura et al. 2012: 9). Bhutan's national approach to wellbeing influenced wider international policies and strategies, including those of UNESCO and UN (UN 2011, UNESCO 2016).

In the last twenty years national and federal governments around the world in countries including the United States, United Kingdom, France, Australia, and Germany have with support from economists developed indices for measuring wellbeing (U.S. Department of Health and Human Services 2000, German Federal Government 2018, ONS 2013, Sharma 2017, DEFRA 2007). These measurements frequently combined objective measures of wellbeing such as income and economic growth alongside subjective measures of wellbeing such as psychological indicators of feelings, emotions, and individual perceptions of satisfaction (Knight et al. 2014). In 2008, the New Economics Foundation (NEF), supported by the UK government, undertook qualitative research into what made people happy and produced five evidence-based actions for wellbeing (Aked et al. 2008). These were: to connect, be active, take notice, keep learning, and give. This research linked wellbeing to both the economy and to wider societal values, and the measurement of NEF wellbeing traits and National Accounts of Wellbeing indicators has been integrated into the wider national wellbeing and population studies, for example by the UK's Office for National Statistics (ONS 2013).

The United Kingdom's approach to wellbeing mirrors that of other national and government agencies, which have measured various quantitative variables, including job satisfaction and the perception that life is meaningful and worthwhile, and formed policies based on these findings. For example, the German Federal Government 'Wellbeing in Germany Report' proposed that wellbeing had twelve dimensions that had a direct impact on 'our' lives (health, education, income, family, and time), on our surroundings (where we live, infrastructure and mobility, security and social cohesion) and on our country (economy and environment, freedom and equality, peace and global responsibility), which could be measured by forty-six quantitative indicators (German Federal Government 2018). These wellbeing dimensions directly correlated to specific government agendas and policies, and subsequently enabled wellbeing elements to be integrated into wider national policies and represent dominant national and political agendas.

The 'utilitarian' economic-based approach to wellbeing enabled it to be integrated within wider national government agendas and policies, which included: economy, education, healthcare, and quality of life (Scott 2012). For example, the US Department of Interior is tasked with preserving, understanding, and showcasing the country's natural and cultural heritage, which aims to strengthen the nation's identity, diversity, and economy (doi.gov/international/what-we-do/heritage). Improving wellbeing was perceived as an important political tool to gain national 'public' support for policies and strategies. The importance of wellbeing politically continued to grow in the 1970s and 1980s, and the emergence of social indicator and quality of life movements in the United States and Europe based in Aristotelian theory meant that by the 1990s wellbeing was increasingly integrated into national policies (Scott 2012: 19).

In 2007, the Department of Education in the United Kingdom included wellbeing education as part of the compulsory primary and secondary school syllabus under the heading 'Personal, Social, Health and Economic

Education in Schools' (Evans 2018: 34–45). In 2009, the UK National Institute for Health and Care Excellence (NICE) suggested that schools should measure the happiness of children in the classroom as a means of identifying any relationship, parent, or family issues (Knight et al. 2014: 1). From the early 2000s, several UK government departments integrated wellbeing into wider policy and government practice. The Department of Health, Department of Education, Treasury, Department for Environment Food and Rural Affairs, and the Department for Environment and Transport and Planning required the appraisal of wellbeing in all their policy documents (DEFRA 2009, DETR 2000, Department of Health 2009, 2010a, 2010b). These national policies have also permeated into local policies and practices, including regional and workplace wellbeing policies, and into specific policies for heritage (Department of Health 2012, Historic England 2022).

Heritage Policy, Theory, Policy, and Wellbeing

Regardless of the awareness of the importance of wellbeing to global citizens and to global development by healthcare professionals, national governments, and international global agencies, the complex matrix of social values linked to wellbeing has until recently been overlooked by many heritage professionals. This is surprising as the social value of heritage has long been established, and for centuries, perhaps even millennia, has been regarded as critical to understanding ourselves and the world around us (Simpson 2010). Heritage, in the physical form including historical sites, archaeological sites, and artefacts, helps people explore and reflect on the human dimensions and conditions of life by transforming these physical objects into stories, memories, and traditions (Simpson 2010). It could be concluded that from the beginnings of human life there has been an innate interest in the past, which created the concept of 'heritage'. The public, as individuals, families, communities, and nations have sought to record, collect, and research the past for the benefit of not only themselves but for the good of the wider public and society (Plato, Aristotle, Rose 2009, Haidt 2007). Heritage was 'saved', collected, stored, and catalogued by individuals not only to obtain new knowledge about the past and to preserve fragments of the past for future generations, but to act as a moral and social tool for the present and for the future; subsequently heritage is intrinsically linked to complex social values and virtues associated with wellbeing (Lowenthal 1998).

From the eighteenth century, the elite assumed that heritage was something that everyone was aware of and as such 'needed' to be saved for the knowledge and social benefit of all (Cleere 1984). From the nineteenth century, academics claimed that it was 'their' moral and ethical duty to protect 'heritage' for the greater good of humanity and that heritage was undeniably valuable to everyone (Cleere 1984). In many countries in Europe, the state took control, on the public's behalf, of the protection and management of this wider social asset, and heritage was controlled by the minority for the benefit of the majority. For example, in the United Kingdom by 1882 the Protection of Ancient Heritage Act was established (Howard 2003, Cleere 1984: 61). The idea was formed that all people implicitly valued or should value their heritage and that heritage and engaging in heritage, for example visiting a museum, was something that the public automatically valued.

The Modernist and Processual Approach

The processual approach of the 1950s had implications for public involvement and engagement with heritage sites (Johnson 1999, Willey and Phillips 1958). According to the processual movement, the protection, interpretation, and presentation of heritage as 'factual evidence' and 'truth' was prioritized and heritage's value was in the provision of and presentation of knowledge and learning (Sayer 2010). This approach based the interpretation of the past on modernist concepts, on 'tangibles' and facts; the socially 'intangible' values of the past were not prioritized. The disciplines of history and archaeology were regarded as professional 'scientific' disciplines, and the past could only be understood through professional and academic training and interpretative technologies that were outside the reach of the 'public' (McGimsey 1972). However there were opposing theories regarding how the public engaged in heritage through the processual/modern 'structural' approach; for example Copeland (2006) described it as enabling simplified and less confusing versions of the past; whilst Hems (2006) suggested that the processual approach produced constructed facts based on official, controlled, and inflexible views of heritage presented as it 'was' in the past rather than in its present context, therefore missing an understanding of present influence and value. Consequently, some regarded the processual approach as separating the public from the past, from connections to 'their past' and 'their heritage', and from developing a sense of ownership, belonging, and custodianship, which in some cases led to the exclusion of the public and communities from heritage (McGimsey 1972, Hems 2006).

The Critical Approach: Post-Modern and Post-Processual Approach

The post-modern, post-processual, and critical approach of the 1980s and 1990s embraced an interpretative methodology based on multiculturalism and acknowledging intangibles and tangibles (Blackburn 2006, Harrison 2013, Sayer 2022). It was claimed that this facilitated the acceptance of external ideas and public involvement in the interpretation of the past, through taking an interdisciplinary approach to the subject, incorporating elements of anthropology and sociology (Binford 1983, Hodder 2001: 1, Harrison 2013, Smith 2004). This approach required heritage professionals to consider values attached to the past beyond that of knowledge and not solely based on the physical remains. A recognition of different interpretations of the past by diverse stakeholders enabled the tangible evidence to be considered in light of the intangible values attached to the past in the present by communities (Munjeri 2006, Harrison 2013). The breadth of definition of heritage that this enabled, both tangible and intangible, initiated a communication between heritage professionals and communities, and enabled indigenous and local communities to have increased involvement in what some regarded as a less politically determined approach to the past (Lowenthal 1985). This approach was, by some professionals, perceived as having had positive effects on heritage, enabling the opening up of heritage to a broader audience (Smith 2004, Hewison 1987). The value of heritage received political recognition and placed heritage and the past firmly in the political and public funding sphere. This was highlighted by government white papers, public speeches, and government support and funding for public-facing heritage activities such as the United Kingdom's Portable Antiquities Scheme in 2006, which has been regarded by some as one of the largest and most successful 'national' community heritage programmes in the United Kingdom (Lammy 2006).

The influence on heritage of this approach was in part the result of a renewed consideration of ethics, including the rights of people to have access to and be part of knowledge production that attached their own diverse values to the past. These principles guided methodology within heritage practice, including the incorporation of consultation, collaboration, community partnerships, and public involvement in heritage projects. This in turn resulted in the growth of the subdisciplines of public and community archaeology, community, and public history; and associated projects, particularly in the United States, United Kingdom, Australia, and Europe (Fredericksen 2002, Greer et al. 2002, Sayer 2014). These principles to heritage practice appear to be directly linked to the indigenous rights movement, which became prominent in both academic and heritage management in the early 1990s (Greer 1995). The discussions related to the rights of the indigenous and/or native 'communities' to be involved in decisions relating to the treatment of 'heritage' on 'their' land (Layton 1989). Despite the United Kingdom and some of Europe having a less obvious and vocal indigenous rights movement, these countries were still impacted by indigenous rights issues and the political movement influenced heritage practice, specifically regarding the inclusion of public views and the incorporation of community consultation within projects.

The critical approach to heritage considered the diverse social values attached to the past and heritage by people (Hewison and Holden 2006, Lowenthal 1985, Harrison 2013). These complex ideas about heritage's social value stemmed from a wide range of disciplines including social anthropology and psychology (Hendry 1999). The complexity of the potential values of the past to the public, specifically social value, was beginning to be reflected in theoretical heritage literature (Binford 1983, Trigger 1989, Hodder 1986, Shanks & Tilley 1987, Smith 2004, Harrison 2013, DeSilvey 2017).

Academics, including Lipe and Darvill, began to undertake detailed analysis of the multitude of social values and attitudes associated and attached to heritage (Lipe 1984, 2007, Darvill 1995, Howard 2003, Thomas, R 2004, Shennan 1988). Lipe's research considered values as a resource and discussed how past values were used in the present for symbolic, informational, aesthetic, and economic reasons, which highlighted how the past's functionality in the present critically impacted the formation of modern-day values (Lipe 1984). He highlighted how values and their interactions with heritage acted as part of a system in which the past was placed in the present. Yet, Lipe did not analyse how different values interact with each other, something that is essential to understanding how heritage affects social values, such as wellbeing (Lipe 2007: 291). Darvill (1995, 1999) divided values into three interacting systems, which included 'use', 'instrumental', and 'institutional'. This value system sought to understand the intrinsic values heritage had to the public, including heritage's impact on individuals' emotions linked to wellbeing (Darvill 1995). For example, how public of engagement with heritage could generate a sense of belonging and connect people to the present (Darvill 1995).

Until the 1990s the social values attached to heritage, including wellbeing, were rarely a consideration for many heritage professionals and policy makers. It was not until 1995 that UNESCO recognized and accepted that 'the intangible had for long been an ignored heritage' (UNESCO 1995: 94) and it took until 2003 for UNESCO to develop and deliver a strategy for safeguarding intangible cultural heritage, and to recognize the importance of social values associated with heritage. Heritage professionals started to deviate away from preserving and presenting tangible physical remains that often centred on its knowledge value, to considering and incorporating the intangible social values attached by the public to the past in the present (Munjeri 2006, Smith 2004: 4). The past's impact on the public and the value the public placed on the past was regarded with social values equal to knowledge values (Munjeri 2006: 334). The 'social' values and subsequent benefits of public and community-centred heritage became increasingly prevalent in academic, professional, and government literature during the early twenty-first century (Howard 2003, Merriman 1991, Council of Europe 2005, Simpson 2009, DeSilvey 2017). The benefits of public involvement in heritage were openly discussed by professionals, academics, and politicians, including heritage's ability to create a 'sense of place' and 'pride' in the place people live, promoting a sense of community, reducing crime, and providing support for those with mental and physical disabilities (Ascherson 2000, Simpson 2010, Isherwood 2009, English Heritage 2004).

Research into values undertaken during the 1990s and the early twenty-first century highlighted the nexus of interwoven values associated with heritage, including social, political, economic, and educational value (Lipe 2007, Darvill 1999, Hewison 2006, Simpson 2010). Heritage value research did not initially provide a mechanism for social value analysis, and as a result these values were often espoused and unquestioned by heritage professionals. The idea of nexuses of interwoven values, the diversity of stakeholder values, and the right for equality of 'stakeholder voices' resulted in conflicting multiple values (Smith 2004, Carman 2016).

Within the heritage profession intangible values had become viewed, in some circumstances, in the same light as beliefs, all of which were deeply personal and should be accepted without question (Blackburn 2006: 4). This posed issues for the profession, as the use of 'evidence'-based interpretation, preservation, and protection declined; the paradigm shift in heritage in some cases encouraged the political manipulation of heritage

(Smith 2004, Smith and Waterton 2012, Simpson 2010, Sayer 2014). The potential social and political value and misuse of heritage and its role in national politics prior to and during this period was highlighted as an ethical issue (Layton 1989, Ucko 1989). Heritage interpretation and presentation could be and was manipulated by the authority of 'those in charge', and despite claims of equality of all stakeholder values, in practice this was rarely upheld; as a result, heritage was often viewed through the lens of 'Authoritative Heritage Discourse' (Smith 2004).

The divergent political and social values associated with heritage at points resulted in stakeholder conflict (Smith and Waterton 2009). The concept of the equality of values led to indecision regarding the prioritization of values and an inability to choose what elements of heritage were important (or were not) to preserve and interpret. This disbelief in absolute values from some professionals resulted in a lack of conviction in heritage management strategies. Subsequently public and professional roles became fluid, and heritage professionals became confused about what heritage was, as well as its wider relevance and impact on modern society. The equality of heritage values and interpretations may have negatively impacted on the public's engagement in heritage and its social impact, with the multiple interpretations confusing and alienating the public rather than educating, informing, and including them (Hems 2006).

The changes to the perception of value did not impact many heritage professionals, as they continued to focus on quantitatively measured economic and educational values (Burtenshaw 2014, Cabe 2004, Jeppson 2008, Jones 1995, 2004, Rizzo and Towse 2002). Social value was perceived by professionals as more difficult to evaluate and quantitatively define, and its evaluation was overshadowed by measurable outcomes. The value of heritage was seen through the political and commercial lenses of cost effectiveness, economic growth, contribution to Gross National Product, and ability to meet external educational needs (Schadla-Hall 1999, Burtenshaw 2014, Rizzo and Towse 2002).

Crosby's work in Fiji and Burtenshaw's work in the United Kingdom highlighted the potential for heritage to increase economic growth through greater work opportunities and a rise in house prices and tourism (Crosby 2002: 362, Cabe 2004, Burtenshaw 2014). Research into heritage values also focused on learning and knowledge, yet this did shift away from the value of formal 'classroom' learning to heritage's impact on informal, practical social learning translated through watching, doing, thinking, and feeling (Henson 2004, Hills 2001, Schadla-Hall 1999, Pearson

2001, Mackenzie and Stone 1990: 4). Research identified that visiting heritage sites supported informal, unscheduled, and impromptu learning and had both learning and social value, with visitors perceiving connecting with heritage as being fun and relaxing (Bahn 1996: 6, Whitaker 1995: 7, Stone 2004: 8). The post-processual and post-modern paradigm supported the identification of informal 'learning' and social values of heritage sites, paving the way for the heritage discipline to consider the impacts of heritage on values associated with wellbeing and happiness, and firmly linking heritage to the NEF five ways to wellbeing framework of learning and connecting (Sayer 2015, Sayer 2018, Aked et al. 2008).

The Future

The transition of heritage practice from modernist (processual approaches) to postmodern (post-processual) critical frameworks encouraged heritage to become a more politically active discipline (Hamilakis and Efthimis 2013: 182, Sayer 2022, Harrison et al. 2020). This changing theoretical backdrop supported the heritage sector's ability to develop political relationships for both professional and wider societal benefit (Harrison et al. 2020). This approach does not advocate that heritage professionals should become political tools, rather that they become politically aware. As such, heritage professionals have connected and applied their work to and within current political agendas, including wellbeing (Thomson et al. 2012, Chatterjee and Camic 2015, Sayer 2014, Sayer 2018, Darvill 2019, Harrison et al. 2020, Sofaer et al. 2021, Brizi et al. 2023, Monkton 2022, Cutler 2022).

The post-modernist paradigm was based on humanistic theoretical and methodological principles. These proposed that individuals could alter the conditions of their existence through action, which has led to suppositions regarding participation in heritage influencing social values, such as wellbeing (Thomas 2004: 1, Sayer 2015: 249). A humanistic perspective consider that humans are interdependent, both desiring autonomy and independence but crave for affiliation and oneness (Pirson 2017: 21). As Pirson indicated, heritage practices that allow for the protection of dignity and the promotion of wellbeing are part of humanistic heritage management (Pirson 2017: 21). This concept has provided a mechanism from which evidence-based heritage research sought to influence wider public policy, in areas such as wellbeing and health care. This has enabled heritage to engage with new disciplines, such as medicine and psychology, in order to provide a robust platform from which to gain financial support for future projects: for example, the Human Henge Project (Darvill et al. 2019). This approach is particularly important in a climate that has seen budget cuts across the cultural sector, hence wider political appeal and demonstrable societal benefit is essential to heritage's professional sustainability (Sayer 2014). As a result, the heritage sector has adapted a postmodernist approach, which has altered practice from one of knowledge acquisition and preservation of the past for future generations, to a societal tool for a wide range of public concerns, including mental health and minority representation (Fujiwara et al. 2014, English Heritage 2006, English Heritage 2014). Heritage has become an ethnographic process, a social activity in which cultural boundaries are overcome and community relationships and collective partnerships are formed; it has been 'decolonized' (Hamilakis and Efthimis 2013: 181).

The application of the humanist approach to heritage practice considered the socio-situational frameworks in which heritage organizations operate, understanding heritage's role and value within its community setting. Heritage projects are viewed as a social practice, which can reflect and contribute to a community's image of itself (English Heritage 2000, 2006, Thomas 2004: 195). Social impacts of heritage to the community are frequently researched and discussed by professionals and academics (Thomas 2004: 1, Lipe 1984, 2007, Darvill 1995, Holtorf 2005, Ascherson 2000, Simpson and Williams 2008, Simpson 2010, Hodder 2000, Kiddey and Schofield 2011, Renfrew 2006, Smith 2006, Kiddey 2017, Gallou and Fouseki 2019). Research into social values reflects changing approaches to value analysis within the sector, with many ideas based on the concept of a nexus of value systems interacting with one another. For example, the past can be perceived as active in the present, impacting on both current and future individual and community values (Simpson 2010). The concept of value networks has enabled heritage professionals to recognize the public value of their work and establish practices that have societal impact (Kiddey and Schofield 2011, Simpson 2010, Sayer and Sayer 2016, Smith 2004, Kiddey 2017, Darvill et al. 2019, Brizi et al. 2023, Gallou and Fouseki 2019).

Heritage professionals and academics have progressively developed an understanding of the multitude of attitudes and values the public place on 'their' heritage and the past, including complex social values (Darvill 1995: 302, Bender 1998, Schadla-Hall 2004, J. Thomas, 2004, Holtorf 2005,

Smith 2006, Simpson 2010, Kiddey 2017, Darvill et al. 2019, Gallou and Fouseki 2019, Harrison et al. 2020, Smith 2021). Understanding of these complex values has enabled professionals and academics to gain insight into the potential impact of heritage on individual and social wellbeing (Sayer 2018, Darvill et al. 2019, Everill et al. 2020, Sofaer et al. 2021, Brizi et al. 2023). The ability to understand the impact of heritage on wellbeing and mental health was support by a shift in theoretical paradigms underlying heritage practice; from 'processual' to 'post-processual' in archaeology, and from a 'modern' to 'post-modern' and even to a 'counter-modernist' framework in history (Sayer 2022). These all represent philosophical shifts in balance and control, ownership, and power, impact, and value (Sayer 2022). This has been supported by political and economic changes and increasing national and global pressures for sustainable futures (Sayer 2022, Harrison et al. 2020).

Yet, until recently, few national or international policy documents have indicated the impact the heritage sector has to wellbeing, and specifically how heritage can support the global societal goal of creating healthier societies. Despite the limited examples of impact research, in the last decade, in Europe and some western countries, including the United Kingdom and New Zealand, wellbeing has filtered into government dialogue pertaining to the methods, practice, future strategic plans, and funding for heritage (New Zealand Government 2019, Historic England 2022).

Heritage includes a diverse range of public participatory and non-participatory activities associated with discovering the past; for example, visiting and engaging in museums, archaeological excavations, and historical sites. In many countries around the world, heritage is a major public pastime, leisure activity, tourist activity, and hobby (Holtorf 2005). As such, heritage, and the consumption of heritage by the public impacts the human, social, and economic capital; and politically the heritage industry is important—it is worth a significant proportion of the Gross National Product of many countries (Sayer 2019: 284). For example, direct and indirect spend from visitors to heritage is worth an estimated 3% of the United Kingdom's GNP (Sayer 2019: 285).

Until more recently, heritage's impact was largely dominated by an association with this GNP, and economic value; yet as previous research suggests heritage can directly impact public and political values, including social values (Simpson 2010, Accenture 2006, Lipe 2007, Smith 2004, Kiddey 2017). Heritage sites and projects perform a role in shaping and supporting values associated with wellbeing, such as the social value of bringing people together, connecting communities and providing a sense of belonging (English Heritage 2000, 2006, Reily et al. 2018, Historic England 2022, Monkton 2022, Gallou 2022). Heritage can play a significant contribution to the way people feel about their lives and the people around them and provides opportunities for improving people's wellbeing. The Council for Europe's Convention for the Value of Cultural Heritage to Society Faro Convention 2004 emphasized the importance of heritage to human rights and democracy, and promotes understanding, relationships, and its use, suggesting that actions and reactions to heritage can result in societal change (Faro 2005). The public's interaction with heritage therefore has the potential to create societal change and impact on wellbeing and mental health (Abdallah 2017, Lakey et al. 2017, European Union 2022). For example, the Understanding Society survey indicated that people who participate less often in heritage related activities have lower life satisfaction and worse physical and mental health, and the What Works Centre for Wellbeing have suggested that 'engaging in heritage activities is associated with lower wellbeing inequality' (Abdallah 2017, Lakey et al. 2017). The concept that heritage can impact wellbeing, and that heritage has intrinsic value to positively impact on wellbeing and make a critical contribution to regeneration is not only the subject of theoretical discussions but is now starting to be academically and professionally researched (DCMS 2016, Historic England 2021, Everill and Burnell 2022, Darvill et al. 2019, Sayer 2018, Brizi et al. 2023).

In the United Kingdom, wider wellbeing agendas and policies have begun to pressure government-funded organizations and government bodies to undertake and fund scoping studies, and to develop heritage-specific strategic plans and funding priorities that incorporate wellbeing. These include Historic England, the National Heritage Lottery Fund, and Natural England, as well as national heritage related charities such the National Trust and Heritage Alliance. In the United Kingdom, a large number of heritage-based wellbeing policies and strategic documents have been influenced by the New Economic Foundation 'Five Ways to Wellbeing' documentation, wider government wellbeing agendas, and academic wellbeing heritage research. For example, Historic England's 'Wellbeing and the Historic Environment' (Reily et al. 2018); Historic England 'Heritage and Society' (2017); Historic England 'Wellbeing and Heritage Strategy' (2022); What Works Centre for Wellbeing 'Heritage and Wellbeing: The Impact of Historic Places and Assets on Community Wellbeing: A Scoping Review' (Pennington et al. 2018); the National

Heritage Lottery Fund 'Wellbeing Framework' (2018); the Heritage Alliance 'Heritage, Health and Wellbeing' (Price and Keynes 2020). To take one example, in 2022 Historic England, the UK government funded heritage advisory agency, launched its Wellbeing and Heritage Strategy (HE 2022). This strategy recognized the importance of heritage in improving people's lives and indicated that by consciously focusing on wellbeing they aimed to make sure everyone experiences the wellbeing benefits of heritage (HE 2022, theheritagealliance.org.uk). The Wellbeing and Heritage strategy was linked to Historic England's Strategy for Inclusion, Diversity, and Equality (IDE) which identified four priority groups to support people to thrive through heritage; this included young people, aging, loneliness, and mental health (HE 2022: 16). Key strategy areas included work, people, and potential, and a holistic, people-centred, collaborative approach that makes a difference was proposed (HE 2022).

In New Zealand, cultural wellbeing has been specifically incorporated into the Local Government Amendment Act, which was passed in 2019 (New Zealand Government 2019). This Act is supportive of the impact that participation in museums and heritage sites (together with the freedom to retain and express cultural heritage) can have on enjoyment and community wellbeing (New Zealand Government 2019). The values of cultural heritage to wellbeing are centred on the Local Government Act 2002 'four wellbeings' and include: social, economic, environmental, and cultural (Ministry for Culture and Heritage 2002). The Act seeks to enable the provision, protection, interpretation, promotion, activities, and access (including participation) to cultural heritage for the good of community wellbeing (New Zealand Government 2019). Government policies and scoping documents pertaining to wellbeing have influenced national heritage funding, practice, and wider disciplinary strategic heritage plans. In some countries, such as New Zealand, heritage wellbeing has permeated broader national and local wellbeing agendas and policies.

The heritage sector's growing interest in its role in supporting community, visitor and society wellbeing has resulted in various advisory bodies being launched, including the Culture, Health, and Wellbeing Alliance (UK), and the Coalition for Arts, Heritage, and Wellbeing (US). These organizations aim to provide a professional network, support advocacy, and provide resources to enable heritage places to support the health and wellbeing of visitors and communities.

In the United Kingdom, heritage organizations, including Historic England, are working with national and local authorities and mental health charities (including the National Health Service and the National Academy for Social Prescribing), to develop mental health and wellbeing projects that are based on participants being referred by doctors and link-workers (Historic England 2022). Heritage social prescription projects have integrated heritage into local wellbeing and mental health policy and practice. Examples include regional and national projects such as the UCL Museums on Prescription Project; Pathways to Wellbeing; Well-City Salisbury; Creative Cultures and Hidden Heritage (Cutler 2022). In Europe, social prescribing and mental health heritage programmes have recently been piloted in countries including Italy, Cyprus, and Greece, for example LOGGIA (Ripanti 2022b). In the United States, heritage organizations, such as Nemours Estate, have developed hospital-based programmes with medical professionals to support the recovery and mental health of patients, relatives and staff (Nemours Foundation 2023).

Heritage-specific and heritage-influenced wellbeing policies have the potential to change the mechanisms heritage professionals use to engage the public in the past, including how to interpret, preserve, present, and promote heritage to the public. For example, in 2022 the European Union's CultureForHealth report indicated that value of the cultural sector in contribute to peoples eudaimonic and hedonic wellbeing capabilities suggested that culture including galleries, museum, and heritage sites should be integral to the EU's health strategy (European Union 2022). However, as noted by European Union, the impact of these policies will not be realised unless there is a collaborative effort by heritage professionals and academia to consistently provide rigorous evidence through research relating to the impact heritage has on wellbeing (European Union 2022). A variety of heritage organizations, primarily in the United Kingdom, have developed strategies for evaluating and capturing the impact of wellbeing, including national government agencies such as the National Heritage Lottery Fund, Historic England, the National Trust, and CADW (the historic environment service of the Welsh Government). Other types of organizations involved include charities such as Waterloo Uncovered and the Restoration Trust, independent heritage sites such as Compton Verney, and museums and archives such as the Museum of London and Archives Plus. Organizational interest in heritage wellbeing is mostly United Kingdom and European-centric at present and does not represent a global trend for assessing the impact of public engagement in heritage on wellbeing. Consequently, this subject area is, although emerging, not comprehensively studied and critically investigated at a global level. For example, there is no cross-asset comparative analysis and no understanding of the role of contextuality of wellbeing and heritage, specifically how national, local, or regional social, political, cultural, and economic climates influence heritage's impact on wellbeing, both of which are addressed in this book.

Summary

International, national, and heritage-specific wellbeing policies have been influenced by trends in wellbeing research. Public policy linked to wellbeing has, in much of the western world, focused until recently on economically based national-centric strategies using quantitative evidence as a proxy for wellbeing (Knight et al. 2014: 19, Scott 2012). As such, policy approaches to wellbeing have been dominated by economists and positivist epistemology, grounded in modern capitalist agendas (Knight et al. 2014: 19). These approaches have failed to comprehend the diversity, plurality, breadth, and potential of wellbeing to impact a wide range of policies beyond economy and healthcare. Current approaches lack the ability to consider people and 'wellbeing' as an individual and social experience, and as such the impact of culture, context, and resources on wellbeing, for example the impact of varying heritages on wellbeing, has been missed (Knight et al. 2014: 26). Recently, global, and non-western concepts, definitions, and measurements of wellbeing, which linked wellbeing to healthcare and social impacts, have permeated into wellbeing research and national policies relating to wellbeing. This change to the breadth of wellbeing-related policy and wellbeing impact can in part can be attributed to international organizations and their adoption of a more holistic globally appropriate definition of wellbeing.

Future wellbeing policies require the development of a new evidence basis situated in humanistic and relativist approaches to wellbeing. This approach requires the consideration that the variables and 'domains' of wellbeing can be measured both quantitatively and qualitatively, and subjectively and objectively. As such, wellbeing measurements need to move beyond economic centricity to incorporate anthropological and ethnographical elements. National and international wellbeing policies need to view wellbeing through this humanistic, contextualized, and individualistic lens, and wellbeing policy should be decentralized to enable actions and strategies that seek to improve wellbeing tailored to the needs and wants of individuals and specific social groups (Knight et al. 2014: 26, Scott 2012, Evans 2018). Critically, wellbeing policies should consider how small locally specific actions can improve wellbeing, including how community resources such as cultural heritage could be utilized to support improved wellbeing and decrease wellbeing poverty and inequality.

This chapter has offered evidence that heritage could positively influence wellbeing, and that public engagement in diverse heritage activities can contribute to general wellbeing policy and has the potential to improve individuals through inclusive practices, including Hedonic and Eudaimonic wellbeing. The heritage sector is theoretically and methodologically lagging behind other cultural industries, and often fails to provide evidence of its impact on wellbeing. The role of heritage has yet to be comprehensively considered, despite the large body of research pertaining to its wider social values. This must be addressed if heritage is to maximize its impact on public health policy and compete with other cultural industries, including theatre and literature. Proactively demonstrating the global potential of heritage to support international health and wellbeing agendas will open up heritage to a new wave of partners, funders, and participants. Proving the value of heritage to society is critical to its future. This book will demonstrate and communicate to a range of stakeholders the global potential of heritage to support international health and wellbeing agendas.

Heritage and Wellbeing: The Impact of Heritage Places on Visitors' Wellbeing. Faye Sayer, Oxford University Press.
 DOI: 10.1093/9780191914539.003.0001

2
Measuring Wellbeing in Heritage

This chapter presents a synopsis of the key evaluation strategies applied to measure the impact of heritage on subjective wellbeing. It outlines the diversity of methodological approaches to 'wellbeing' evaluations undertaken by heritage professionals and academics regarding museums, archaeology, and heritage's impact on individual social values related to wellbeing. This research has included quantitative and qualitative research from a variety of contexts, albeit mostly in the United Kingdom. This chapter outlines the research strategies undertaken for different types of heritage assets, specifically museums, archaeological excavations, and heritage sites.

The chapter considers the use of quantitative and qualitative evaluation to capture evidence of heritage's impacts on wellbeing. Through critical discussion it outlines the merits and issues of applying different methodologies of data capture to measure wellbeing and identifies a robust methodology to assess the impact of heritage on wellbeing. It evaluates the feasibility of the heritage sector presenting high-quality evidence to demonstrate its wider national and international role in enhancing wellbeing and outlines the research strategy applied to the case studies presented in 'Heritage and Wellbeing'. This strategy is based on pre-approved public health measures and pre-tested anthropological and ethnographical approaches. The methodology underpinning the research presented in 'Heritage and Wellbeing' involves a double helix 'mixed method' approach, combining pre-tested qualitative and quantitative measures of subjective wellbeing, including the Positive and Negative Affect Schedule (PANAS) and Modified Wellbeing Scale (MWS). The research methodology presents a global framework for analysing heritage wellbeing that can be applied in a range of cultural contexts and offers guidelines for future wellbeing-based evaluation frameworks for the cultural heritage sector.

The Importance of Evaluation

Evaluation, judging the importance of something, specifically, its value and impact, is essential for understanding human behaviour, actions, and reactions. Social evaluation can provide a baseline of data to demonstrate current patterns in human behaviour and can support and justify the development of strategies and interventions. Evaluation enables the evidence-based identification of gaps in current provisions and strategies to be identified and future plans to be formed. Evaluation of specific projects and programmes provides evidence of impact and value and presents data that can demonstrate change and determine whether specific interventions have had a positive or negative impact. At social evaluation's core is the ability to self-reflect, to learn from previous actions, and adapt in the future.

Evaluation can take different formats that seek to obtain different or comparable data sets. These evaluations include:

- *Planning*: Pre-project or event evaluation collects baseline data from consultation and previous experience and can feed into the project planning and design. Pre-project planning evaluation is a public-centred approach (Belcher 1991). It acts as form of market research of stakeholders and has become standard practice in project planning (Merriman 1991, Sayer 2018). Evaluation techniques can involve focus groups with potential stakeholders to receive feedback on new initiatives and ideas. Planning evaluation enables projects to adapt and innovate and to mitigate the risk of failure.
- *Formative*: This, often informal, evaluation is done during the project by the project team. It aims to test the project's value, identify gaps in provisions or elements that are and are not working, and enable the project to adapt and change before completion. This can include participant comments and feedback, acting as a form of product testing.
- *Summative*: After the project has been completed, summative evaluation provides formal feedback as to the actual values and impacts of the project. This method can offer impartial evaluation of projects if it is undertaken by those outside the project. It aims to evaluate lessons learnt and provide ideas for future plans.

Formats for evaluation use specific methodological practices, either qualitative, quantitative, or a combination of the two. Quantitative surveys provide statistical data, which can indicate visitor trends to inform future practice (Woollard, 2004, Merriman 1991, Belcher 1991, Hooper-Greenhill 2004). These often take the form of visitor surveys, which provide demographic information about the visitor, time spent, and routes of movement. This type of analysis gathers empirical data through closed-question interviews and multiple-choice questionnaires. This produces quantitative information that can be subjected to statistical analysis, such as Merriman's museum visitor survey (Merriman 1991). The 'quantitative' approach is based on questions that guide the participants into producing limited answers, which are usually focused on market values (Merriman 1991). The results from quantitative surveys provide numerical estimates of the ideas of the person who designed the survey and can simplify complex concepts, whilst qualitative methodologies draw on anthropological and sociological methods to enable the broader analysis of social and personal values (Sayer 2018). This approach produces specific individual responses rather than numbers and generalizations (Sayer 2018). Qualitative evaluation uses various techniques such as visitor conversations and observations. It can involve analysing comment cards, recording open interviews, or observing participant reaction and actions, including visitor footfall and interaction.

Policy makers often favour quantitative evaluation and numerical scales to test and assess overarching patterns in the behaviour of population and the impact of policies: this has included the measurement of the wellbeing and mental health of individuals, communities, and nations (ONS 2013, 2018, Evans et al. 2015). In part, this relates to ease of administration, as self-assessment surveys, often online, can be easily administered to large populations, with low costs, at multiple points in time. Such surveys offer longitudinal evidence of change and large data sets that can be analysed quickly using computer programmes. Numerical scales provide easily understandable data sets that can readily be translated into 'fact' and illustrated visually in tables and graphs to provide evidence of the quantitative impact of policies and interventions and can inform future practice.

National population surveys have recently integrated indicators of wellbeing and mental health within their numerically based scales to capture information. Examples include the UK government's Office of National Statistics national survey (ONS 2013, 2018, Evans et al. 2015). National surveys can act in much the same way as a pain measurement plan, for example, 'over last few weeks how you have felt on scale of one to five regarding…'. Museums are regarded as trail blazers in applying a 'government'-applicable quantitative approach to demonstrating the impact of activities on wellbeing and mental health (Thomson and Chatterjee 2013). Mixed methods offer scope for understanding wellbeing; potentially they offer a more valid mechanism for understanding the impact of heritage on wellbeing. This chapter will consider the variety of methods used in multidisciplinary arenas, considering each method's merits and issues. It will outline the evaluation methodology used in this study based on the analysis of these methods and an understanding of the heritage context and integration of both disciplinary and government needs in terms of baseline data required to act and implement change.

Evaluation of Heritage and Wellbeing

Evidence for health, wellbeing, and quality of life depend on the interactions between physical, psychological, and social functions (Paddon et al. 2014). For this reason, the production of an evidence base within the cultural sector, including heritage, has been a challenging task for professionals and academics. Despite the complexities of measuring wellbeing and mental health, often identified by intangible social values and individual experiences, the heritage industry has started to trial mechanisms to evaluate and provide data to assess the evidence previously regarded as inaccessible and to support hypotheses of heritage's positive impact on wellbeing.

Heritage organizations and researchers have used a variety of quantitative, qualitative, and mixed methodologies to access the impact and value of heritage on wellbeing. The majority of these have focused on qualitative evaluation methods to identify wellbeing and mental health improvement markers (Kiddey 2017, Power and Smyth 2016, Finnegan 2016, Johnston 2019, Clifton 2019). This qualitative evaluation, often in the form of informal participant feedback, has required little changes from normative methods for understanding the outcomes of visits to specific heritage places and engagement in heritage projects. Few studies have sought to identify measurable change to social and personal or subjective and psychological wellbeing resulting in participation and engagement in heritage.

The identification of tangible and intangible values attached to heritage relates to the complexities of the evaluation of social values in individual communities (Munjeri 2006). Holtorf suggested that heritage values are not merely group values and nor are they purely externally determined, but they are created within a specific community and by the individual (Holtorf 2005, 2006). They can only be deconstructed by using qualitative methods, which enable an understanding of the value of heritage to the individual. Yet, a large proportion of work relating to wellbeing and heritage has principally focused on the identification of the possible outcomes and not considered how these could be rigorously analysed (Hewison and Holden 2006). Heritage professionals have suggested that the analysis of heritage's ability to change longer-term values, including wellbeing, should incorporate externally created models for evaluation, including public service models and those applied by healthcare professionals (Accenture 2006).

To date, museums and galleries have set the standards for producing high-quality research into wellbeing through combining quantitative and qualitative evaluation to understand projects and wider sector impact on wellbeing (Thomson et al. 2012, Smiraglia 2015, Chatterjee and Camic 2015: 183–186, Kindleysides and Biglands 2015, Pennington et al. 2018). Meanwhile, the evaluation of heritage and archaeological sites has, in most cases, focused on qualitative surveys using individual case studies, often not explicitly based on deconstructing their impact on wellbeing (McDavid 2012, Atalay 2007, Simpson 2010, Belford 2011, Kiddey 2017). Exceptions to the individual case study-based approach include Sayer's quantitative research on the impact of community archaeology excavations on wellbeing (Sayer 2015) and Sofaer et al.'s mixed method UK-based research into the subjective and psychological wellbeing effects of visiting heritage sites after Covid (Sofaer et al. 2021). As a result, the wider discipline of heritage is not being consistently evaluated through rigorous evaluation strategies and appropriate analysis. The heritage sector is failing to engage with or develop appropriate evaluation methodologies to enable it to 'prove' its impact on wellbeing. This chapter seeks to address this imbalance, providing an overview of measurements of wellbeing that can be applied to evaluate heritage's impact on wellbeing.

Qualitative Evaluation

Research into the complex and often intangible social values attached to heritage, including those linked to wellbeing, have primarily used qualitative social science methods (Jones 2004, Simpson 2010). It has been asserted that quantitative methods, those that produced numbers such as demographic analysis, visitor numbers and closed question survey could not provide an understanding or evidence of 'intangible' personal and social values (Merriman 1991, Jones 2004, Simpson 2010, Whelan 2015). The 'social' impact of heritage was until recently focused on qualitative methods applying a cross disciplinary approach utilizing anthropological, sociological, and ethnographic methodologies (Jones 2004). Consequently the mechanisms used to investigate the social values of heritage have focused on participant observation and in-depth open qualitative interviews (Jones 2004, McClanahon 2006, Ripanti 2022a). The application of anthropological methodology has supported the deconstruction of the multiple values and the contextual investigation of social values attached to heritage (Bender 1998).

Qualitative methodologies, which drew on anthropological and sociological epistemological frameworks, enabled the analysis of social and personal values (Emerson, Fretz, and Shaw 1995: 143, Bryman 2012, Sayer 2018). These methodologies measured individual responses rather than numerical generalizations (Sayer 2018, Bryman 2012). Qualitative studies have used a range of methodologies to assess the myriad social benefits of heritage, which have included:

1. *Consultation Workshops/Focus Groups*: These are structured around providing the stakeholders, including the public, with the opportunity to discuss ideas and values and share thoughts about specific projects. Workshops aim to provide an open dialogue between either a specific or broad population demographic depending on the intended audience of project. This methodology can act as market research, uncovering and discovering communities' agendas, wants and desired outcomes, or actual outcomes of a project and can act as catalysts to establish collaboration and active engagement.

   Workshops and focus groups can incorporate multiple evaluation techniques, including semi-structured questions, observations, conversations, self-reflection, and feedback walls. They provide opportunities to develop 'genetic social outcomes' through assessment of social values and help to define outcomes and establish effective evaluative frameworks (Whelan 2015: 220). Pre-project National Lottery Heritage Fund (NLHF) evaluation at Compton Verney (UK) used consultation workshops and focus groups to establish goals, objectives and outcomes and shape the activity plan. At the National Lottery Community Funded Well-City Salisbury, pre-programme stakeholder workshops were used to develop the project's theory of change model based on impact of heritage on wellbeing (Fig. 2.1).
2. *Open Questionnaires and Surveys*: These are loosely based around quantitative surveys and can be delivered in person or remotely through post, email, or online. Qualitative survey questions are

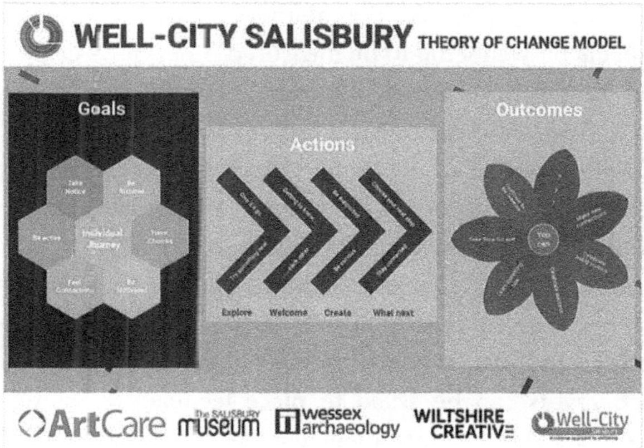

Fig. 2.1 Theory of change model for heritage and wellbeing project Well-City Salisbury (Sayer and Luck 2022).

based around key themes and collected data is analysed in relation to these themes. Surveys can support personal commentary and investigate experiences and motivating factors of stakeholders on specific projects or project elements. Survey questions are open ended, non-leading, and enable answers to be given in free comment space rather than numerical deciphers. Qualitative open-ended questionnaire-based surveys have been used to assess the positive and negative effects of involvement in heritage (Power and Smyth 2016). Questionnaires, including Power and Smyth's survey of NHLF funded heritage landscape project, have sought to capture the motivations, personal journeys, and types of involvement of specific social groups and assess if involvement in heritage can promote wellbeing (Power and Smyth 2016).

3. *Semi-Structured Interviews*: Semi-structured interviews utilize techniques developed from oral history and are based around recorded individual or group interviews. Interviews are informal and conversational, utilizing a semi-structured thematic approach to enable an organic and flexible conversational style. These interviews are based around themes or categories, and facilitate the interviewee to set their own agenda, in a relaxed environment that supports people to be open and honest (Jones 2004: 8). The relationship between the interviewer and the interviewee can influence the information that is received and its relevance.

This method enabled the nuances of social values and their impacts on wellbeing to be measured. Interviews can be linked to consultation, pre-project planning, and the workshop (Simpson 2010, 2018). This method has been used by Nimenko and Simpson, and Finnegan, to access how archaeology can aid the recovery and skill developments of injured veterans and impacted their wellbeing (Nimenko and Simpson 2014, Finnegan 2016).

4. *Feedback Walls, Comment Cards, and Comment Space*: These surveys require individuals to write down their thoughts and feelings and can be based on pre-determined questions or ideas. Participants can be asked to place feedback or comment cards within a pre-determined themed area—a 'feedback wall'. Conversely feedback through this method can be collected without the use of thematic prompts; this free commentary can involve drawing pictures or writing. Cards can be added to a comment wall or placed in box. Recently heritage projects such as Oakington 'Bones without Barriers Project' have used social media, such as Facebook and Twitter, to collect comment-based social impact feedback based on comments and messages from project pages (Sayer and Sayer 2016, Sayer 2018).
5. *Self-Reflection*: Self-reflection captures data, often informally, from material created by participants in specific projects. This method can include data such as written reflections, drawings, photographic evidence or video logbooks, vlogs, and diaries. Analysis employs a 'reflective method' using anthropological techniques to understand the values and perceptions of the impact specific activities, projects, and interventions (Hodder 2000). The anthropological framework enables analysis to consider the wider social, political, cultural, and historic contexts (Hodder 2000). Self-reflection can facilitate individuals to be introspective about their experiences and through self-awareness analyse their own personal changes in values as a result of involvement in heritage. This method offers scope for understanding how heritage influences values from the viewpoint of those involved in each stage of the project. This approach has been used by Kindleysides and Biglands at Beamish Museum to support the assessment of mental health and wellbeing impacts to individuals involved in specific heritage-based activities, and by Clifton at Human Henge to determine the use of landscape for wellbeing (Kindleysides and

Biglands 2015: 274, Clifton 2019). It is worth noting that self-reflection can require individuals to keep up to date with their activity logs and engage in the process.

6. *Observation*: Observational analysis involves watching and recording the actions of the public, participants, and professionals, including conversations, behaviour, body language, and personal interactions in a specific context or during a specific time, or on a specific project. Observations are recorded digitally, or through notes, photographs, or videos; observations from the project staff can also be collected. This technique is based in techniques used in anthropology and ethnography and involves looking for changes in behaviour and listening informally to conversations between people and changing speech to identify changes in social values, including those linked to wellbeing. Observational analysis requires thematic analysis, an understanding of social norms, behavioural patterns, and human behaviour. This enables analysis of what people think and how they interact, seeing the users 'public' as active rather than passive receivers. Observational analysis in heritage settings is seldom used alone, rather it is combined with other elements including conversational and dialogue analysis and self-reflection. This approach has been used by Sayer and Sayer on the 'Bones Without Barriers Project' and Kiddey on the 'Homeless Heritage Project' to investigate social impacts of heritage, specifically including mental health and wellbeing (Sayer and Sayer 2016, Kiddey 2017).

Qualitative ethnographically situated evaluation can provide evidence of the impact of heritage projects on wellbeing. Qualitative measurements have frequently been applied to investigate the value of heritage to wellbeing and mental health (Sayer 2018, Power and Smyth 2017, Ripanti 2022a, Finneran and Welch 2022, Johnston 2019, Clifton 2019, Darvill 2019). The methodological flexibility and organic nature of qualitative measurements of wellbeing enables it to be adapted to a variety of socio-political, economic, and geographical contexts (Kiddey and Schofield 2011, Sayer and Sayer 2016, Darvill 2019). The broad thematic scope and extensive data collected from qualitative surveys means that pre-existing 'social value' data from previous heritage evaluations can be re-examined, enabling new inferences to be drawn including impacts to wellbeing (Reily et al. 2018).

Quantitative Evaluation

Recently, heritage professionals and academics have recognized that qualitative data alone is detrimental to data collection and analysis (Merriman 2004, Thomson et al. 2012, Sayer 2018). The use of qualitative data alone can inhibit heritage's ability to provide externally 'politically' relevant and usable evidence of its social impact (Chatterjee and Camic 2015). This is particularly poignant within the mental health and wellbeing arena, which requires evidence based on rigorous statistical analysis, using accredited public heath evaluative methods to ascertain impact. Subsequently, heritage evaluation has begun to apply public health perspectives to assess impact on wellbeing. Using methods that are accepted and approved by government agencies and health care professionals enables heritage to incorporate pre-existing evaluative frameworks and therefore potentially influence policy relating to wellbeing and to heritage itself (Chatterjee and Camic 2015).

Quantitative surveys collect statistical data, which can indicate trends to help inform future practice (Reeve and Woollard 2007, Sayer 2018). Frequently these take the form of visitor surveys, which provide demographic information about the visitor, time spent, and routes of movement. This type of analysis gathers empirical data through closed-question interviews and multiple-choice questionnaires and produces quantitative measurement that can be subjected to statistical analysis (Merriman 1991). The quantitative approach is based on questions that guide the participants into producing selected answers (Newman 1995: 321). This method develops questionnaires centred around multiple-choice questions, which can be used to measure specific tangible and subjective impacts of heritage projects (Hooper-Greenhill 2004). The results from quantitative surveys provide numerical estimates of the ideas of the person who designed the survey.

The heritage sector has applied quantitative frameworks to measure wellbeing, using psychologically accredited measures, such as Positive and Negative Affect Schedule (PANAS) and EuroQol Health Status Visual Analogue Scale (VAS), to measure the impact of specific projects of wellbeing (Thomson et al. 2011, Sayer 2015). A variety of quantitative wellbeing measurement frameworks have been applied with heritage, including:

1. *Self-Reported Surveys*: Measured by self-scoring descriptive statements or questions on a Likert scale. The Likert scale asks participants to rank answers to questions from range from 1–5 (e.g. 1 = not at all, ranging to 5 = all of the time). The numerical scales can be adapted from 1–5, 1–7, 1–10, or 1–00, which enables these raw figures be transposed into percentages. Questions can either be subjective or objective and, for example, subjective questions might include which feeling: cheerful, relaxed, active, rested, and interested. These questions can be administered before and after activities or over different time periods to look for change. A modified version of the WHO-5 survey was applied on the World Heritage Site Lijiang to investigate the factors which influenced residents' subjective wellbeing using a 1–7 scale (Chi, Cai, and Li 2017). Thirty-four subjective wellbeing questions were linked to economics, social relationships, sense of community, social environment, positive tourism, and negative tourism, and used to investigate cognitive and overarching indicators of individual wellbeing.
2. *The Warwick-Edinburgh Mental Wellbeing Scale (WEMWBS)*: This has provided the building blocks for numerous heritage wellbeing surveys (Tennant et al. 2007, Thomson et al. 2011, Sayer 2015, 2018, Heaslip 2019). For example, the UCL museum wellbeing survey used a twelve-item generic VAS wellbeing self-answered questionnaire. Participants were asked to circle statements to indicate levels of agreement, based on a 1 (none of the time) to 5 (all of the time) Likert scale. Statements included: I feel happy, I felt engaged, I felt comfortable, I felt safe and secure, I enjoyed the company of other people, I talked to other people, I was interested, I enjoyed it, I was amazed, I was entertained, I felt confident, I felt healthy (Thomson and Chatterjee 2013). The 'Can Digging Make You Happy Project' utilized modified elements of the WEMWBS VAS (MWS). The questions were adjusted to incorporate the NEF five wellbeing elements and applied to a 1–10 Likert-type scale of measurement (Sayer 2015). These questions included: Thinking about yourself, how interested are you in the world around you? Thinking about your own life, at present how connected do you feel to people around you? When considering your personal happiness, at the moment how happy would you rate yourself?

Thinking about your own life and personal circumstances, how satisfied are you with your life as a whole?

3. *The Generalized Anxiety Disorder Questionnaire (GAD-7)*: A self-administered Likert scale-based questionnaire (Kroenke and Williams 2006, Kroenke et al. 2010). The questions are based on indicators for anxiety and designed to identify anxiety-related problems (Kroenke and Williams 2006). Participants are asked to answer seven questions by assigning a numerical judgement from 0 to 3 for how they felt over the preceding two weeks. The scores can be added together, and an overall score used to provide an indicator of individual levels of anxiety and statistical significance testing can be applied (Kroenke and Williams 2006, Kroenke et al. 2010). GAD-7 has been used alongside other measures as an indicator of improved veterans' mental health and wellbeing as a result of involvement in heritage activities and to research relating to links between participation in creative heritage activities, wellbeing, and anxiety (Nimenko and Simpson 2014, Everill et al. 2020).
4. *Patient Health Questionnaire (PHQ)*: Used to diagnose mental health disorders in adults, such as depression and anxiety (Pfizer 2017). It asks patients a number of self-reported questions based on a scale of 0–3 (0 = not at all, 3 = nearly every day). There are specific measures within this, including PHQ-9 (depression), which is a measure for depression and has been adapted to test change in veterans' wellbeing and mental health as a result of heritage-based interventions (Nimenko and Simpson 2014, Everill et al. 2020). As with GAD-7, by adding scores together high scores are considered indicators of depressions or mental health disorders.
5. *The Visual Analogue Scale (VAS)*: The most frequently used global method for testing the impact of museums on visitors or the community. The popularity of VAS relates to its pre-approved and pre-tested multidisciplinary approach to investigating values, as well as the practicalities and ease of data capture. VAS is easily administered either in person or through online questionnaires, Apps, and other digital platforms, for example the Mappiness project, which surveyed over 1 million people's feelings related to happiness and associations with a variety of heritage sites through the development of an App (Fujiwara and Mackerron 2015). Its methodological frameworks enable questionnaires to be adapted to investigate specific values that

were appropriate, understandable, and suitable for diverse demographics and apply to a variety of geographical, economic, social, political, and cultural contexts.

Heritage organizations including Manchester Museum, the National Museum of Wales, Holburne Museums, and Wessex Archaeology have recently employed external agencies and specialists to administer VAS-based wellbeing surveys (Garcia and Winn 2017, Sayer and Luck 2022), Externally contracted wellbeing evaluation has been used to investigate sustained improvement in wellbeing and has been placed within multiple wellbeing measurement frameworks, including VAS, with surveys based on WEMWBS and National Accounts of Wellbeing Research, Quality Adjusted Life Year (QALY), and Social Return on Investment (SROI) measurement (Garcia and Winn 2017). In the United Kingdom, a number of national heritage organizations including Historic England, the National Heritage Lottery Fund, and the National Trust have contracted the 'What Works Centre for Wellbeing' to create specific surveys to measure heritage's impact on wellbeing, all of which have utilized VAS methodological frameworks and adapted WHO and WEMWBS indicators (Reily et al. 2018: 5, Pennington et al. 2018).

6. *The Positive and Negative Affect Schedule (PANAS)*: PANAS was developed in 1988 by Watson, Clark, and Telligen to measure positive and negative moods (Watson et al. 1988). PANAS, unlike WHO 5 and WEMWBS, was designed to look for both positive and negative affects to wellbeing. The central component of PANAS are words linked to positive and negative emotions measured by self-scoring on a 1–5 Likert scale: (1) Very slightly or not at all, (2) A little, (3) Moderately, (4) Quite a bit, and (5) Extremely (Watson et al. 1988). Positive words included: attentive, interested, alert, excited, enthusiastic, strong, inspired, active, proud, and determined. Negative words included: distressed jittery, guilty, afraid, irritable, ashamed, scared, hostile, nervous, and upset (Watson et al. 1988). Measurement of wellbeing is based on participant scores, for example if a participant gives a low score on positive words and a high score on negative words, they would be regarded as unhappy. PANAS is usually delivered before and after engagement in specific activities to generate two measurements, to facilitate longitudinal analysis.

PANAS is a complex method for investigating subjective wellbeing, and as a result is used less frequently than VAS-based

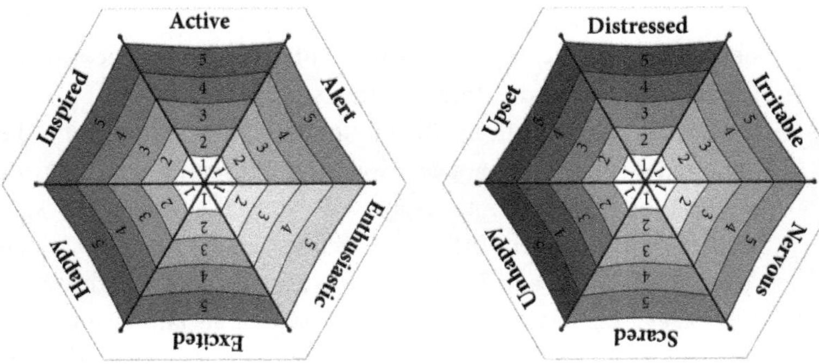

Fig. 2.2 UCL museum wellbeing survey 'The Umbrella' (Thomson and Chatterjee 2013).

methods. PANAS has been adapted to investigate the positive and negative impacts of heritage activities on wellbeing and has included: museums' impact on wellbeing; community archaeological excavations impact on wellbeing; investigating both positive and negative emotional impacts of specific projects on wellbeing; and the impact of World Heritage sites on residents (Thomson et al. 2011, Thomson and Chatterjee 2014, Sayer 2015, Chi, Cai, and Li 2017).

The UCL museum survey was designed around a colour-coded visual aid of an umbrella, on which participants circled scores 1–5 based on strength of feeling within the umbrella for each word (Fig. 2.2, Thomson and Chatterjee 2013, Thomson and Chatterjee 2004, Thomson and Chatterjee 2015). Simplified positive words included happy, inspired, excited, alert, and enthusiastic, and the negative words included upset, distressed, unhappy, sacred, nervous, and irritable (Thomson and Chatterjee 2013, Thomson et al. 2011). The survey was moderated for specific age groups, for example positive words for younger people included: motivated, lively, positive, talkative, friendly, and interested; whilst for older adults positive words included: cheerful, encouraged, enlightened, inspired, absorbed, and active. Comment space was provided to capture more detailed information for change and improvement (Thomson and Chatterjee 2013).

Mixed Method Evaluation

The mixed methodology approach to evaluation combines quantitative and qualitative approaches to collect multiple data sets, such as PANAS,

VAS, conversational data, open-ended question surveys, interview data, and behavioural analysis data (Nimenko and Simpson 2014, Sayer and Sayer 2016, Sofaer et al. 2021). The mixed method approach produces data patterns that enables researchers to map how specific heritage practices and actions are associated with specific behavioural changes and how these actions impact personal and community wellbeing (Sayer and Sayer 2016). The mixed method 'double helix' approach has been used to understand how heritage can change complex social dynamics and how these dynamics are linked to positive and negative changes in wellbeing (Nimenko and Simpson 2014, Finnegan 2016, Kindleysides and Bigland 2015, Sayer and Sayer 2016, Sayer 2018, Thompson et al. 2011, Sofaer et al. 2021). Thomson et al., Thomson and Chatterjee, and Sayer have all combined PANAS, Likert-type visual analogue scales, free space commentary, and quantitative measures that have been used alongside recording participant, teacher, facilitator, and staff commentary, and personal feedback (Thomson et al. 2011, Thomson and Chatterjee 2015, Sayer 2015, Sayer 2018). These mixed method approaches can provide nuanced wellbeing data and highlight the impact of complex social dynamics, individual perspectives, and values on heritage's effect on wellbeing (Sayer 2018).

The mixed method approach can support the investigation of the impact of specific actions on individual wellbeing-associated 'values'. It can account for context and external factors including social dynamics and individual circumstances during data analysis (Sayer 2018). This provides critical 'value' and 'social impact' data for understanding the complexities of heritage's impact on wellbeing.

1. *The Delphi Method*: A structured method that can combine rounds or numerous stages of stakeholder workshops, focus groups, and surveys, which can be based on open and closed questions to enable participants to reflect, reconsider, and adapt ideas and values to come to a consensus (Keeney et al. 2010). For example, the Amphora project used a mixed method Delphi approach to develop guidelines for involving people with mental health challenges in a heritage project (Solent University 2022). Amphora used surveys based on open and closed questions to come to an expert- and experience-based consensus on the impact and responsibilities of heritage projects on participant mental health. The project used three rounds of questions, which moved from broader closed and open questions in the first round, closed questions in the second round, and closed and expanded commentary

in the third round to come to an agreement on the impact and responsibilities of heritage projects that aimed to support wellbeing (Solent University 2022).

2. *Social Return on Investment (SROI)*: A mixed-method approach that combines qualitative, quantitative, and financial details to enable the critical analysis of the impact and socio-economic value of heritage projects. This method specifically seeks to investigate the social, economic, and environmental impact of specific actions to stakeholders (2011, 2014). It enables the investigation of how projects contribute to objective and subjective socio-economic wellbeing. SROI is a positivist framework that seeks to quantify the unquantifiable, such as self-confidence and friendship, and to understand the breadth of values (Whelan 2015: 222). SROI has been applied as an evaluation strategy in a variety of heritage settings, including at the Human Henge and Manchester Museum (Drysdale 2019; Garcia and Winn 2017). The Inspiring Futures Project used SROI to develop quantitative economically based questions, such as volunteering and staff hours, and costs of time, and qualitative-based wellbeing questions based on 'social' values to volunteers and venues. This method assessed the project's value for money and determined the economic, environmental, and stakeholder impact of the investment (Garcia and Winn 2017: 6).

   SROI is methodologically adaptable and can incorporate qualitative methods within surveys. It provides a varied, contextual, flexible framework for investigating wellbeing. However, it is not without issues, including judgement and discretion in setting, quantifiable value, and agreement over the best way to report results (Whelan 2015: 225).

3. *Self-Identified Aims Wheel (SIAW)*: Developed by Well-City Salisbury and their evaluation consultants to measure change in participant social and personal wellbeing as result of taking part in heritage-based wellbeing programmes. At the beginning of a course or project participants were asked to identify three personal aims they would like to achieve during the programme and write these aims in spaces provided around the outside of a wheel. Participants assigned a numerical judgement 1–5 (1 = setting out to achieve your aim, 2 = making positive steps, 3 = halfway, 4 = almost there, and 5 = reach your aim) on how they felt at that moment in time (Fig. 2.3). At the end of the course or project

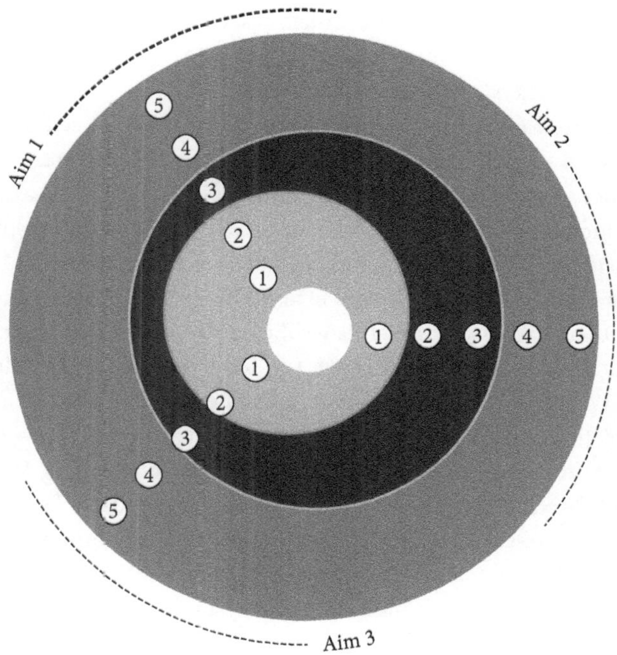

Fig. 2.3 Self-identified aims wheel (Sayer and Luck 2022).

participants were asked to use same aims as identified in the beginning, to assign a numerical judgement on same 1–5 Likert scale. SIAW quantitatively and qualitatively identified thematic patterns and frequency in wellbeing aims chosen by participants and identified significant measurable change in personal aim scores before and after taking part in programme, course, or project, linking this to changes in personal and social wellbeing indicators.

Issues

Wellbeing evaluation methodologies are not without issues in their application to heritage settings, both in their implementation and the type of data they collect. Quantitative surveys do not allow for open-ended responses, or the specific detail required to assess the multiple, unquantifiable, and interwoven personal wellbeing values associated with heritage (Sayer 2015, McClanahon 2006: 127). On the contrary, a quantitative approach can accentuate predetermined research biases and affect the formulation of questions, enabling participants to provide the answers to only the researchers' pre-selected questions.

The majority of heritage wellbeing surveys require the self-assessment by individual participants of subjective wellbeing; this requires emotional intelligence, mental awareness, and honesty. Wellbeing self-assessment surveys including WEMWBS, PHQ, and GAD-7 are often inappropriate for use with children under 16 and for individuals with emotional, behavioural, or social difficulties (Spitzer et al. 1999, Rathouse 2019). The World Health Organization states that its wellbeing survey is appropriate for use with children over the age of 9 (WHO-5, Topp et al. 2015). The Museums' Toolkit Wellbeing survey sought to mitigate the emotionally complex nature of the questions to enable it to be used with children—this was achieved through modifying, simplifying, and colour coding VAS questions and PANAS words (Thomson and Chatterjee 2013). Despite these modifications researchers still identified issues with questions being too cognitively challenging for assessing children's wellbeing (Thomas and Chatterjee 2013). Sayer noted that some adults struggled with self-assessment of wellbeing due to the requirement for high levels of emotional awareness and 'intelligence' to honestly answer questions based on feelings and emotions (Sayer 2015). Many wellbeing-based questionnaires, including those based around VAS such as ONS, rely on participants' accurate memories and individuals' self-assessment emotions over a two-week period (ONS 2013, 2018, Evans et al. 2015, Chi, Cai, and Li 2017). Researchers have noted issues obtaining permissions for the completion of both the before and after surveys by all individuals, and biases towards ranking with preference given to higher numbers and ranking of 4–5 or 9–10 on Likert scales (Thomson and Chatterjee 2013, Thomson et al. 2011, Sayer 2015).

Heritage professionals report that wellbeing results are impacted by external factors, including social contexts and group dynamics, which are often not identified through quantitative methods (Sayer 2015). Measurement frameworks for heritage wellbeing have attempted to account for external factors by asking control questions within surveys, for example: 'In the past two weeks, are there any major events happening in your life that will influence your affective experience?'; 'In the past two weeks, are there any major events happening in your life that will influence your overall life satisfaction?' (Chi, Cai, and Li 2017: 220). Waterloo Uncovered undertook 'planning' evaluation of participants to measure how individuals' backgrounds and external factors influenced individual wellbeing and mental health prior to involvement in heritage projects. These are two examples of how, through qualitative interviews and open-ended surveys, heritage projects have sought to provide evidence of change and understand individual wellbeing journeys (Waterloo Uncovered 2019).

The influence of external factors on wellbeing can be mitigated against by combining qualitative and quantitative methodologies, including observations, conversations, self-reflections, and free commentary. Free commentary space provides useful contextual and personal feeling-based data, although this commentary can be impersonal (Kindleysides and Biglands 2015). The application of wellbeing-based evaluation requires ethical consideration, including of how approaches might alter participant responses and behaviour, and increase social and individual stresses and tensions, as well as potentially causing emotional distress.

The analysis of data can be affected by researcher biases, for example thematic biases in qualitative analysis and statistical biases in quantitative analysis. Many heritage wellbeing surveys have not measured the negative effects of heritage on wellbeing, with the exception of PANAS-based surveys (Sayer 2015, Thomson et al. 2012, Reily et al. 2018). Quantitative analysis of the data is often based on conversion of data to simplistic percentages and percentage comparisons, in part due to having insufficient data to undertake significance testing, for example at Human Henge (Heaslip 2019). Such a quantitative approach can fail to investigate statistical significance and consider that any change might not be statistically significant. The majority of heritage wellbeing surveys do not engage in longitudinal analysis and therefore cannot measure change in wellbeing as a result of specific heritage actions. Such non-longitudinal surveys provide a baseline data of wellbeing of an individual that might not be related to involvement in heritage (Nimenko and Simpson 2014).

The vast majority of issues with the application of wellbeing measurements within heritage settings can be mitigated against in the planning stages and through the use of a mixed methodology. Mitigation of issues requires demographic and contextual adaptation of questions and the application of appropriate statistical and thematic analysis. To effectively measure heritage's impact, the heritage professional must carefully consider the planning evaluation of the project, and learn from others, taking advice from multidisciplinary advisors and wellbeing specialists on appropriate measurements and analysis.

Analysis

Analysis of data provides evidence of change, yet the analysis of raw data is complex as data is often multifaceted and if longitudinal can be collected from different evaluation stages. Analysis will depend on the type and level of detail required, and the nature of data collected, for example psychological or subjective wellbeing or social, human, or economic wellbeing capital.

Quantitative Analysis

The implementation of statistical analysis on quantitative results investigates overarching data patterns and enables comparative analysis between different data sets. Descriptive statistics involves comparing percentage differences between the mean results from different data sets. This enables differences and variations to be identified between two sets of data, aiding comparative analysis from before and after involvement in a heritage project, between different demographics or social groups, or between different types of heritage resources, to name a few examples (Sayer 2015). Advice on 'raw data' analysis of museum wellbeing surveys has suggested that key to comparative analysis is working out the average mean and medium and finding a range; the difference between the first and second results (before and after) (Thomson et al. 2011). This produces results that can be used to look for changes in wellbeing over time, for example if a participant gives a score of 4 before and 3 after then the range is −1, which would suggest lower levels of wellbeing after participation in heritage.

Mean, medium, and mode results can be converted into percentages to create average scores for groups or individuals over time. Percentages enable comparative analysis of data, an assessment of change, and variation between sets of results. WHO suggests that following data collection, all the raw data scores from the five categories can be totalled (total scores will range from 0 to 25) and multiplied by four in order to create a percentage (Topp et al. 2015). The subsequent percentage is regarded as a representation of wellbeing, with a higher percentage being equated to higher wellbeing (WHO 1998, Topp et al. 2015). A score out of 100 enables results to be converted into percentages and changes to be measured in percentages, with significant differences highlighted; for example, 10 per cent change between scores would be considered significant (Topp et al. 2015).

Complex statistical analysis can require the use of specialist computer software such as SPSS (Statistical Package for the Social Sciences), enabling the researcher to engage in robust statistical analysis, including significance tests such as standard deviation and paired sample tests. Standard deviation determines the significance of results; the lower variability between results illustrates patterns in behaviour and abnormal variance that should be excluded as insignificant. Complex data analysis has been applied to determine longitudinal and group wellbeing change and differences between data sets, for example the Wilcoxon Paired Sample test and Mann-Whitney U test, which have been used in research by Thomson et al. evaluating the therapeutic effects of museum object handling with hospital patients, and Sayer's community archaeological research (Thomson et al. 2011, Sayer 2015, 2018).

The Wilcoxon Paired Sample test compares longitudinal results (before and after scores) from the same group. The Wilcoxon Paired Sample significance test has been implemented on heritage wellbeing data to test for improvement, not just difference (Sayer 2015). The Wilcoxon test is used on pairs of related data sets, for example data from prior to, during, and after visits, and specifically looks for change. The dependent variable is measured at a continuous level, for example MWS 1–10-point scale and PANAS 5-point scale. The independent variable consists of two categorical, related groups; for example the same individual participants from the same project are measured at two intervals, prior to and after the project (Sayer 2015).

The Mann-Whitney U significance test compares groups and identifies statistical differences. It has been used to compare two means from two independent populations (Thomson and Chatterjee 2013, Thomson et al. 2011). This approach is valid when the dependent variable is ordinal or continuous, as is the case in Likert scale tests, and when the independent variable consists of two unrelated and categorically different types of groups and types of projects, for example gender, age, and type of heritage sites (Laerd Statistics 2018). Analysis requires different participants in each group and the same number of participants in each group. If these conditions cannot be met then Wilcoxon is a more appropriate form of analysis.

The Paired Sample Sign Test is non-parametric and is used to determine if the mean difference between two data sets from the same individuals or groups of individuals is significant. This can be used as a longitudinal test, to look for change in individual visitor scores before and after taking part in heritage projects or visiting sites. As with Wilcoxon the dependent variable should be measured at a continuous level, for example a 1–5-point Likert scale. The independent variable should consist of two categorical, related groups, for example the same individual participants from same project measured at two intervals, prior to and after the project.

Qualitative Analysis

Quantitative data analysis of written, verbal, and visual material (e.g. transcriptions of oral data, visitor commentary, written self-reflections, observational notes, photography, and video diaries) requires the investigation of patterns in words, visuals, and behaviour. A requirement of qualitative analysis is a constant dependable variable, for example all participants interviewed took part in a community archaeology excavation, and an independent variable, for example community archaeology excavations are in different geographical contexts (Sayer 2010).

Qualitative analysis requires the production of a pre-project evaluative framework that is based on key questions and continuous and consistent data collection. Analysis searches for answers to key questions based on constructivist theories and requires the thematic of analysis of material. The process of thematic analysis attempts to discover patterns in social values and behaviour based on key criteria 'themes', for example specific changes to categories of wellbeing values based on engagement with heritage (Jones 2004, Simpson 2010). The identification of patterns in data can be based on a grounded theory, which aims to systematically highlight themes, factors, experiences, and motivations (Strauss and Corbin 1990, Power and Smyth 2016). The data will be grouped into categories to look for overarching themes and patterns in data. This form of analysis translates individual responses into group thematic patterns based on specific criteria and enables the investigation of impact. Thematic approaches to qualitative data include various forms of analysis dependent on medium.

Content analysis studies of documents and communication through the presence of certain words, themes, and concepts. This qualitative approach enables researchers to systematically quantify and analyse the presence of certain patterns in these themes, such as in visitor comment cards and in visitor conversations. An example is the Eichstedt and Small's study of systematic evaluation and analysis of rhetorical presentation strategies relating to enslavement in contemporary museums. It identified themes in dominant narratives and motivation for dominant narrative and counter narratives and identified areas of variability (Eichstedt and Small 2002). Critical Discourse Analysis analyses text, vocal content, conversation, or any communication to explore the meaning behind communication (Wodak 2012). As part of this analysis language is regarded as a social practice, in which knowledge is constructed and produced (Smith 2011). Discourse analysis seeks to uncover dominant hegemonic discourse and understand mechanisms used in heritage setting that could support views or values being modified, for example wellbeing values attached to museums (Smith 2006).

Other approaches to qualitative analysis include integrated Problem Tree Analysis or situational analysis, which are often used in focus groups to understand drivers, goals, and issues; group analysis and interpretive phenomenological analysis, which explore how people, as individuals in a particular context, makes sense of a particular phenomenon in their 'lived experience'; and narrative analysis, which focuses on how people use stories to make sense of the world. All of these have been used as methods to collect and measure wellbeing data (Centre for Thriving Places 2016).

Mixed Method Analysis

Mixed method analysis combines a qualitative and quantitative analytical approach to triangulate and validate results. This form of analysis is based on grounded theory, and themes might also be identified in pilot quantitative or qualitative research or by pre-ascribed theme, for example the NEF five ways to wellbeing.

Factor analysis enables a large number of observed variables. This might be for example within visitor commentary or open question surveys, to be reduced into a smaller several variables through correlation of these into larger underlying themes (Finch 2020). This might include wellbeing themes such as power, place, and people (Pennington et al. 2018, Ripanti 2022a). Factor analysis has been used by Sofaer et al. in the 'Places of Joy' project, combining data from qualitative and quantitative evaluation to measurably and thematically identify core themes and subthemes in the perceived improvements to hedonic and eudaimonic wellbeing of visitors to heritage sites; core themes included capacity, social connections, ontological security, and trust (Sofaer et al. 2021).

Conclusion

The methods outlined in this chapter provide an overview of the normative mechanisms for measuring and analysing wellbeing in relation to heritage. The application of both quantitative and qualitative methods produces data to support the understanding of subjective wellbeing and evidence of change in wellbeing because of specific heritage-related interventions or actions. Yet none of these methods is free from issues; some fail to understand external contributing factors, others struggle to provide evidence of change, some rely on the honesty and emotional intelligence of participants, and many rely heavily on anecdotal evidence translated into 'fact'.

The methods used for evaluation of wellbeing and heritage vary in their ability to provide high-quality evidence of wellbeing impact. The analysis of the multitude of methods used to assess wellbeing impact, many of which have been pre-tested in heritage settings, offers insights into the significance of combining quantitative and qualitative approaches within evaluative frameworks. A 'double helix' model using a mixed method approach to data collection and analysis enables professionals to produce more robust research data to prove heritage's impact on wellbeing. This chapter suggests the investigation of heritage's impact of wellbeing should combine pre-tested multidisciplinary evaluative methods to maximize data collecting and evaluating opportunities to provide depth of understanding and support validity of results. This approach to wellbeing evaluation enables researchers to tailor surveys for use in a variety of heritage environments and socio-political and geographical contexts (Thomson et al. 2011, Aked et al. 2008, Watson et al. 1988, Sayer and Sayer 2016, Sayer 2015, Sayer 2018, Sofaer et al. 2021).

Heritage and Wellbeing: The Impact of Heritage Places on Visitors' Wellbeing. Faye Sayer, Oxford University Press.
 DOI: 10.1093/9780191914539.003.0002

3
Heritage Places and Wellbeing
Methodological Approach

The research used a mixed method approach, which combined qualitative and quantitative methodologies to provide robust, rigorous, and significant evidence of the role heritage places had in enhancing individual visitors' wellbeing. This study aimed to identify significant changes to visitors' subjective hedonic and eudaimonic wellbeing after their visit to heritage places, by surveying visitors at the beginning and end of their visits. Visitor surveys combined closed question quantitative data collection, which included Positive and Negative Affect Schedule (PANAS) and Modified Wellbeing Scale (MWS), two systems of measurement specifically designed to assess changes in subjective wellbeing, including to personal happiness, together with open question qualitative information from the comment space in visitor questionnaires. Through collaborating with heritage places in the United Kingdom, United States, and India, before and after visitor surveys were undertaken in different contexts and sites. This pre- and post-visit data was compared to identify change and the results were analysed, against the diverse types of heritage places, in different locations, and with different visitor demographics. The analysis of quantitative and qualitative data assessed whether it is possible to identify and link changes in individual wellbeing because of visiting heritage places.

This research was carried out to meet a pressing need for heritage places to provide evidence of their impacts on wellbeing. At the same time, heritage places are seeking to understand their social and human capital, and their potential impacts to national wellbeing, health, and sustainable development agendas. The research reflected these requirements and organizational pressures to answer the following questions:

1. To what extent do heritage sites impact visitor wellbeing?
2. Do specific management practices at heritage places impact visitor wellbeing?

3. How can incorporating new measurements relating to personal and social wellbeing provide evidence of the role of heritage sites in supporting aspects of human rights, for example equality?

To answer these questions the research was designed to achieve the following five objectives:

1. Develop and implement a robust, user friendly, and professionally applicable evaluation strategy for investigating the impact of heritage sites on visitors' subjective wellbeing.
2. Identify if public visits to and engagement with heritage places have different impacts on personal and social wellbeing dependent on type of place, location, and visitor demographics.
3. Demonstrate the distinctive methodological practices that make heritage places a significant global resource for positively transforming wellbeing.
4. Develop, through evidence-based research, a globally applicable and sustainable public interpretation/engagement strategy that benefits individuals, community, and societies and seeks to support capacity building.
5. Provide evidence to place heritage sites firmly within national and international priority areas/strategies relating to wellbeing, inclusivity, diversity, and equality.

The heritage sector struggles to cohesively measure its impact on individuals' social and personal wellbeing, even though it has hypothesized it might have potential positive impact, including benefits related to ethical practice, human rights, identity, equality, and inclusivity (Hodder 2010). There are no standardised measures to assess the impact heritage sites have on wellbeing, beyond numerous measures associated with general wellbeing. This means that they are unable to provide evidence of the complex and nuanced emotional reactions in relation to the feelings of wellbeing that visitors experience. As a result, heritage places' engagement and interpretation strategies have either failed to consider how to integrate these strategies to support wellbeing, or they are unsure how this might be achieved. An additional question to be answered is whether heritage sites can neutralise or have a positive impact on aspects of wellbeing associated with historic trauma.

Government heritage agencies and organizations have identified a deficit in sector-wide robust empirical and statistically significant evidence relating to the acknowledged impact that heritage sites can have on visitor and community wellbeing (Reily et al. 2018, Historic England 2017, Pennington et al. 2018, National Heritage Lottery Fund 2018). They have indicated that to have application beyond the heritage sector, including for mental health provision (e.g. social prescribing) and wider sustainability agendas, there is a requirement to apply public health care and psychologically accredited wellbeing measurements that are statistically significant, as well as longitudinal analyses of change to wellbeing that can be linked directly to engaging in heritage sites (Michaelson et al. 2009). Yet measuring wellbeing traits within the heritage sector has not been consistently, strategically, or systematically studied, to assess how wellbeing is influenced by management or interpretation strategies (Historic England 2017, 2022, Sayer 2015, 2018). That said, wellbeing is a central theme in political discussions and is at the epicentre of many public policy documents, including the National Health Service (NHS) (Knight et al. 2014, DEFRA 2009, DETR 2000, Department of Health 2009, 2010a, 2010b, United Nations 2011, UNESCO 2016, WHO 2008, European Commission 2008). Wellbeing agendas have filtered into specific and non-specific health and wellbeing policies, including Historic England's Heritage and Wellbeing Strategy (Historic England 2022). Furthermore, wellbeing, lowering wellbeing inequity, and reducing wellness poverty is a political priority for many international and national organizations and governments (United Nations 2011, World Health Organization and Commission on the Social Determinants of Health 2008, Office for National Statistics 2013, 2018).

As noted in previous chapters, wellbeing (as attributed to a combination of physical, social, and psychological variables) has been measured in public health through the self-reported individual emotional reactions and judgements related to subjective wellbeing, happiness, life satisfaction, and quality of life (Michaelson et al. 2009, Aked et al. 2008). The New Economics Foundation (NEF) five evidence-based actions for wellbeing (to connect, be active, take notice, keep learning, and give) link wellbeing to both the economy and to wider societal values (Aked et al. 2008). For heritage places to support positive wellbeing and be part of mental health provision they must create and support conditions in which these attributes can thrive.

Furthermore, heritage sites are driven by economic management models that do not always support sustainable development within communities. They have not considered how localized experience of heritage and alternative people-centred perspectives could positively influence a public experience, develop community wellbeing, and support global sustainable development agendas (Harrison et al. 2020). As a result, the sector is often not able to support or positively affect change in their communities.

Case Studies

In partnership with heritage sites, archaeological sites, museums, and galleries in the United Kingdom, United States, and India, quantitative and qualitative data collection and evaluation of visitor wellbeing as self-reported by using questionnaires was undertaken from 2021–2022 (Fig. 3.1). A combination of in-person and online visitor surveys were undertaken at study sites, based on public healthcare and psychologically accredited measures. A total of ten sites took part in the visitor wellbeing survey, and which resulted in the collection of robust empirical data of the change in wellbeing after visiting heritage places (Appendix, Fig. 1).

Data collection was originally due to begin in 2020, but Covid-19 and related public health measures such as lockdowns, social distancing, and travel restrictions measures, including the closure of heritage assets, resulted in data collection starting in autumn 2020 and finishing in June 2022. The aim was that at each of the ten case study sites collected one hundred visitor surveys to enable a large enough sample size for statistical analysis to have a high enough confidence rating, to be representative of visitor demographics, and to present findings that could provide generalized conclusions that represented wider population behaviour regarding the changes to visitors' wellbeing following their visits to heritage sites. Due to concerns regarding withdrawals and incomplete surveys, the project was adapted to collect as many surveys as possible from each site within a specific data collection timescale.

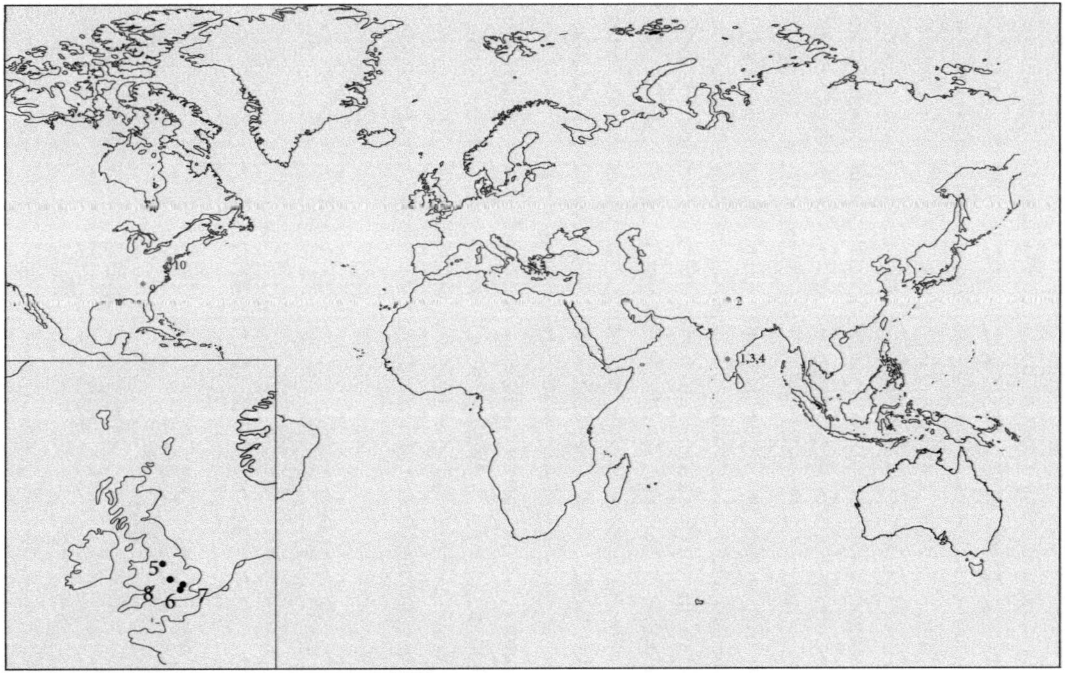

India
1. Bangalore Fort
2. Chandigahr Museum and Art Gallery
3. Tipu Sultan's Summer Palace
4. Venkatappa Art Gallery

United Kingdom
5. Chatsworth House
6. Dulwich Picture Gallery
7. Imperial War Museum Duxford
8. Priory Gardens

United States
9. Biltmore Estate
10. Nemours Estate

Fig. 3.1 Location map of case study heritage places.

Methodological Approach

Public mental health professionals have designed, implemented, and employed evidence-based strategies to provide measurable results for mental health and wellbeing programmes, yet evidence of the unique impacts of heritage places and projects are lacking in part due to skills, time, and the appropriateness of applying industry-based standard measurements within heritage settings. Psychologically robust empirical research into the hypothesized transformative impacts of heritage sites on wellbeing is economically, socially, and politically critical and timely to the profession, including the role of heritage within mental health provision and wellbeing associated with sustainable development policies and agendas. This research tackles this knowledge and evidence gap not only by utilizing published and piloted industry standard measures for change in psychological wellbeing (Positive and Negative Affect Schedule—PANAS) but by developing a heritage specific wellbeing impact measure, the 'Modified Wellbeing Scale'—MWS—and using free comment thematic open questions based on key personal and social wellbeing indicators (Michaelson et al. 2009, Aked et al. 2008, Watson et al. 1988).

Key indicators for improved wellbeing were based on NEF's five evidence-based actions for wellbeing, which are: to connect; be active; take notice; keep learning and give (Aked et al. 2008, Fig. 3.2). Survey questions were generated based on measuring the ability of heritage places

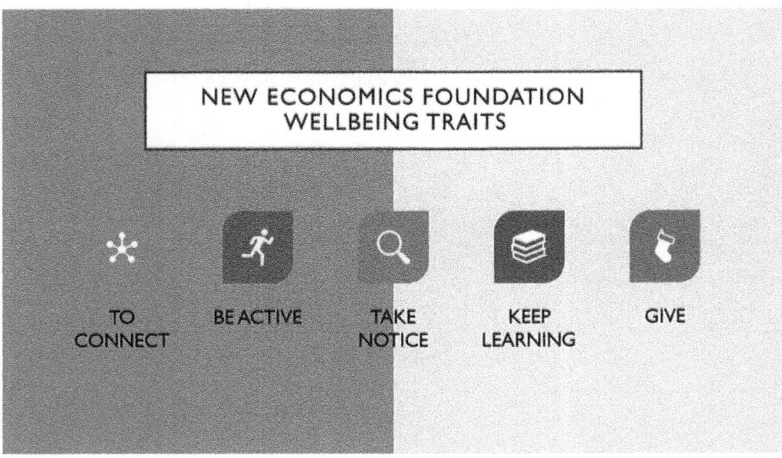

Fig. 3.2 Five ways to wellbeing (based on Aked et al. 2008).

to generate a sense of belonging, to connect people to each other and their local environment, to act as mechanisms for personal and communal interaction with one another and with their environment, to support learning, and to be physically and mentally active.

Additional indicators for determining the impact of visiting heritage place on visitor wellbeing were identified from the National Accounts of Wellbeing (NAW) personal and social indicator framework. These included social wellbeing indicator questions (supportive relationships, and trust and belonging), personal wellbeing indicator questions based on the five components (emotional wellbeing, satisfying life, vitality, self-esteem and resilience, positive functioning) and respective sub-components (Michaelson et al. 2009, Fig. 3.3).

Modified Wellbeing Scale (MWS)

This utilized modified elements of the General Health Questionnaire, the ONS Wellbeing Survey and the Warwick–Edinburgh mental wellbeing scale and incorporated the NEF five wellbeing key elements and NAW personal and social wellbeing indicator framework (Aked et al. 2008, Michaelson et al. 2009). A Modified Wellbeing Survey was designed that integrates questions from psychologically accredited surveys for measuring eudaimonic and hedonic wellbeing, including WEMWBS, along with questions that incorporate the NEF five evidence-based actions for wellbeing and NAW personal and social wellbeing indicators (Aked et al. 2008, EuroQol Group 1990, Evans et al. 2015, Goldberg and Hillier 1979, Tennant et al. 2007, Michaelson et al. 2009, Watson et al. 1988).

Questions in the MWS corresponded to specific personal or social wellbeing indicators and indicator components as identified by NAW (Fig. 3.3, Aked et al. 2008, Michaelson et al. 2009). Indicator questions are based on the capability approach, which associates specific characters to eudaimonic and hedonic subjective wellbeing (Scott 2012: 22, Fig. 3.3, and Appendix, Fig. 3). MWS used a simplified 1–5 Likert-type scale, with questions adapted to ask how visitors felt at the specific moment the questionnaires were delivered to them, as opposed to a broader time range as used by GHQ, WEMWBS, and ONS, and to be used specifically in a heritage place setting (Appendix, Fig. 2, Kroenke and Williams 2006, Tennant et al. 2007, ONS 2018). If the visitors provided higher scores after their visit, their subjective wellbeing would be deemed to have improved. This approach has been applied internationally to a range of heritage projects

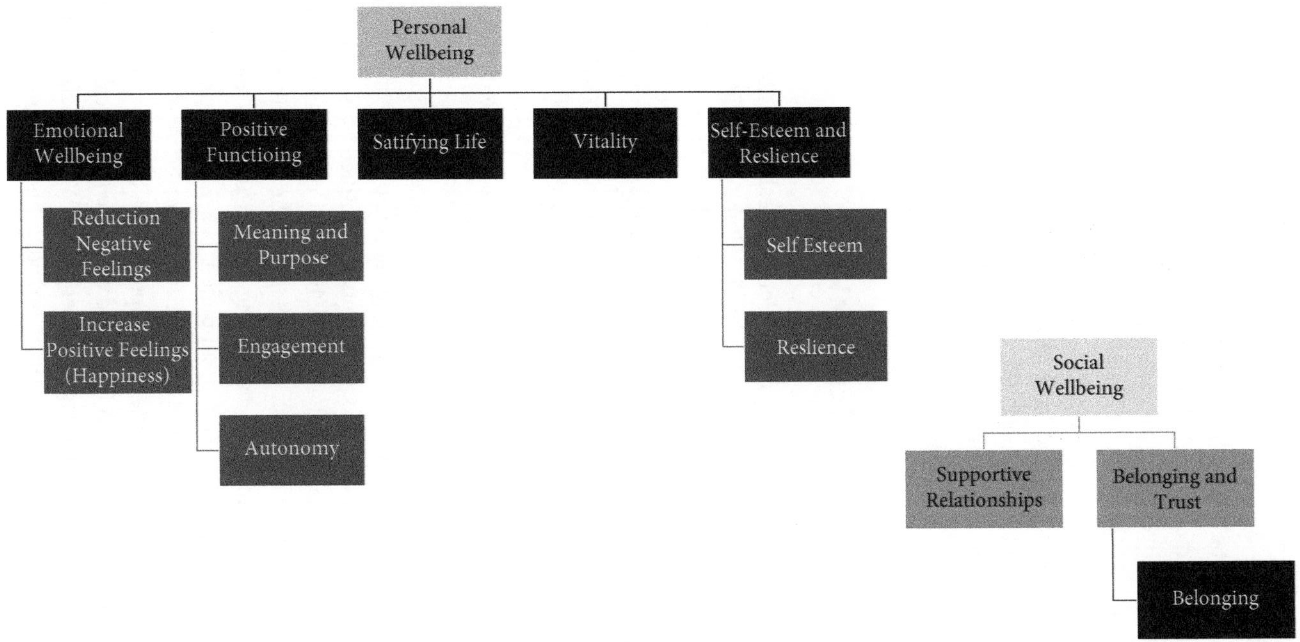

Fig. 3.3 National accounts of wellbeing framework indicator structure (Michaelson et al. 2009).

and places by the author (Sayer 2015, 2018) and enabled the collection of robust and statistically significant evidence that health care, heritage, and policy professionals require to co-produce strategies utilizing heritage sites in social prescribing programmes and mental health provision.

Positive and Negative Affect Schedule (PANAS)

PANAS a public health care accredited measure specifically developed to identify change in personal wellbeing through positive and negative emotions, measured through scoring positive and negative words (Appendix, Fig. 4, Watson et al. 1988). PANAS measures subjective hedonic wellbeing through individual assessment of happiness linked to positive (and negative) effects, and it asks people to rate their feelings on a Likert scale (Watson et al. 1988, Scott 2012: 26). PANAS can also be used to measure changes in specific capabilities linked to eudaimonic wellbeing, including personal indictors and indicator components such as positive functioning, vitality, and self-esteem and resilience (Watson et al. 1988, Michaelson et al. 2009, Scott 2012, Appendix, Fig. 5). Developed to measure change in positive and negative moods, specific words are linked to positive and negative emotions measured by individual choice against a 1–5 Likert scale (Watson et al. 1988). The participant chooses a number from one to five for each emotive word on the list: positive words included attentive, interested, alert, excited, enthusiastic, strong, inspired, active, proud, determined; negative words included distressed, jittery, guilty, afraid, irritable, ashamed, scared, hostile, nervous, and upset (Fig. 3.3). Positive and negative emotions are scored; if a participant scores a low score on positive words and a high score on negative words, they are deemed to have lower personal and emotional wellbeing (Watson et al. 1988).

Visitor Commentary: Open Questions

Qualitative methods included the provision of free space for visitor comments at the end of the survey, based on experience, and 'prompt' questions based on feelings, confidence, and skills, linked to the NAW personal and social wellbeing indicators (Michaelson et al. 2009). Questions included:

- Do you feel the visit impacted your wellbeing? If so how?
- What was the most meaningful experience during your visit?
- Free comment space.

The visitor survey also collected demographic data to provide information required to investigate whether age, gender, ethnicity, and visitor disability impacted how heritage places change social and personal wellbeing indicators. This also enabled a consideration of heritage places' potential role in tackling wellbeing inequality. Data collection was undertaken in person at heritage places by a survey team, or surveys were conducted online and linked to ticket sales. Visitors to heritage places were asked in person or via an email to complete pre- and post-visitor surveys at the start (before) and end (after) of their visit to heritage places. Qualitative and quantitative data collected on visitors' surveys was triangulated to investigate and measure how visitor key personal and social wellbeing changed after visiting heritage places and provided measurable and statistically significant data of changes in individual wellbeing.

Observational analysis of visitor practices and experiences at all case study sites was undertaken by the team. Observations recorded visitor behaviour, including visitor interactions with each other, heritage place staff, and with interpretative material. These observations were documented in self-reflective journal and provided valuable contextual and behavioural information to inform conclusions.

Inclusion and Exclusion Criteria

Visitors to case study heritage places were invited to participate in the study either through email or in person. At the start of the evaluation process the project was explained and participants invited to read a Participant Information Sheet (PIS) and invited to ask questions before taking part, either in person or online. Once participants had read the PIS, they were directed to complete a consent form and were not able to progress to survey questions without its completion.

In all cases, capacity to consent was assumed unless the researcher noticed signs that the participant might not be able to understand the purpose and nature of the research, to understand what the research involves, to understand risks and burdens, to retain the information long enough to make an effective decision, to have free choice, and to make a decision within a specific time frame, for example prior to entering a heritage place (Appendix, Fig. 6)

To increase the reliability of the findings, surveys were conducted at different days (both weekdays and weekends at each site) and different times, to reach as many demographics as possible. Gender, ethnicity, and disability were not factors considered in selecting participants, and neither was completion of the demographic section of the survey. Once the data had been collected a participant coding system was used to anonymize the data, which was stored securely and in compliance with GDPR rules. The research was funded by British Academy small grants (SRG1819\190,822), derived from the Academy's partnership with the Department for Business, Energy, and Industrial Strategy, the Chittick Fund, and the Albert Reckitt Fund. Ethical approval for research was granted by Manchester Metropolitan University and follow up work by the University of Birmingham.

Impacts on Data Collection

Covid-19 and subsequent social distancing, lockdowns, travel restrictions, and institution travel bans resulted in delays to the data collection and in some cases made it impossible to collect data from sites outside the United Kingdom. In Nigeria and Palestine, the author was only able to collect data from a limited number of visitors, which did not provide large enough data sets for statistically significant analysis. These case studies were regrettably removed from this study. Data collection from in-door heritage places, such as museums, was hampered due to institutional health and safety requirements and reduction in visitor numbers, and data from museums in the United States could not be collected during the research period.

Covid-19 required the adaption of data collection methods at heritage places with large internal spaces and high visitor numbers, such as Biltmore Estate and Chatsworth, required due to these organizations' social distancing measures and health and safety considerations. This necessitated the use of online surveys. Online before and after surveys were linked to tickets sales, with the before survey linked to an automatic email sent to visitors twenty-four hours prior to arrival and the after survey linked to an automatic email sent to visitors immediately after their visit. At other case study places where the data was collected in person by the author, research assistant, and/or by trained volunteer and student interns, to meet health and safety requirements the collection was either conducted in outdoor spaces, or the collectors wore masks indoors.

At all heritage case study places there was a high level of automatic withdrawals of data due to incomplete surveys, and failure to complete after surveys in India and the United Kingdom, and in the United States due to the high levels of incomplete before surveys. As a result, participant numbers decreased from 3,749 collected to 1,366 usable. The quality of data collected varied from site to site, for example in Indian sites data analysis was made complicated by high levels of incomplete or partially complete surveys. This meant that these sites had to be withdrawn from the research. Online data collection resulted in higher numbers of visitor surveys being completed, although it should also be noted that this was at Biltmore and Chatsworth Estates which have the largest number of visitors of the sites studied. On the other hand, the use of online surveys also resulted in higher levels of dropouts and required a longer period of data collection.

At Indian heritage places surveys were provided to visitors in three languages, Kannada, Hindi, and English; the majority were completed in Hindi, which were then translated. Most visitors to Indian heritage places did not complete the demographic parts of the survey, and of those that did the majority were under 30. As a result, it was not possible to complete a demographic analysis for Indian heritage places. Data collected at UK and US heritage places indicated a lack of respondent diversity, and as a result there was not enough data to undertake comparative statistical analysis of ethnicity.

Examination and Exploration of Data

Analysis requires visitors' before and after surveys and scores to be matched. If visitors did not complete both the before and after survey, then they were excluded from the data analysis. As a result, the 3,749 before surveys collected was reduced to only 1,366 visitors who provided usable data.

Quantitative analysis was conducted on PANAS and MWS results. Percentage differences between mean results from before and after visits were compared to enable an understanding of overarching data patterns. Statistical hypothesis testing was conducted to determine the significance of changes to individuals' wellbeing after visiting heritage places, and the probability that any change was not a result of chance and showed a true reflection of the target population. The Pair-Sample Sign Test, a non-parametric related sample test, was used. In all tests, the null hypothesis assumed that the median of differences between specific measures (for example 'attentive') pre-test and post-test equalled 0. Results were deemed to be statistically significant when the p-value was below 0.05 (5%) and highly significant below 0.01 (1%). The two-tailed test was conducted to test for change to wellbeing, rather than improvement or reduction. The direction of change was determined by the greater number of positive or negative differences observed in matched pairs. If there were more positive matched pairs than negative, for example, the direction of change was determined to be positive.

Standard deviation from the mean was calculated to ascertain variability. Standard deviation demonstrated how far away on average each answer was from the mean average. Standard deviation differences between the 'before' data and the 'after' data were calculated to assess whether variability had increased or decreased after visiting a heritage place. Lower variability between results was deemed to illustrate patterns of behaviour. If a change in specific score (e.g. 'life is worthwhile') was both significant and variability decreased, then this was deemed to represent a measurable and statistically significant change in a related wellbeing indicator as a result of visiting a heritage place.

Qualitative data from visitor commentary was thematically analysed based on grounded theory. This was supported by NVIVO software to pick out key words and themes. In-depth qualitative evaluation was undertaken by hand by the author or RA, based on colour coding of visitor comments on themes linked to the NAW indicators and the NEF five ways to wellbeing (Michaelson et al. 2009, Aked et al. 2008). Patterns in visitor comments were associated with personal and social wellbeing themes and provide nuanced value and impact data that enabled interpretation of quantitative data. When triangulated, data sets provided self-reflective and personal detail from visitors to start to understand why certain heritage places had impacts to specific wellbeing indicators and categories.

Summary of Approach

The methodological approach used in this research provided quantitative and qualitative data to enable analysis that could triangulate results and provide a more nuanced understanding of what specific attributes of visitor personal and social wellbeing that heritage places could influence. The breadth of case studies, diverse visitor experiences, contexts, and type of heritage supported the analysis of differences between types of heritage places and how the different contexts of heritage places affected their impact on wellbeing. The demographic information provided evidence of variation in the wellbeing impacts of heritage sites based on gender, age, and whether the visitor had a disability.

Heritage and Wellbeing: The Impact of Heritage Places on Visitors' Wellbeing. Faye Sayer, Oxford University Press.
 DOI: 10.1093/9780191914539.003.0003

4
Heritage Sites and Wellbeing

This chapter examines the impact of heritage sites, especially historic houses, historic estates and buildings, and archaeological sites on visitor subjective wellbeing. It will explore the hypothesis that physical engagement in heritage sites, through visiting and participating in experiences within these sites, including storied experiences and interpretative narratives, can benefit visitors' subjective well-being. Through thematic and statistical analysis of quantitative and qualitative data collected from six heritage sites in the United States, United Kingdom, and India, research explored how visiting heritage sites impacts on visitors' personal and social wellbeing at both a contextual and sector-wide level. Through comparative analysis differences and overarching patterns in data between diverse heritage sites, this chapter identified the specific attribute of wellbeing heritage sites changed, and the strategies and practices at these heritage sites which positively impact visitor wellbeing.

Heritage Sites and Wellbeing

The foundations for research into the potential of heritage sites (including monuments, historic buildings, and historic landscapes) to impact on wellbeing have for the most part come from landscape geography or ecology (Tacon and Baker 2019, Forestry Commission 2013, Cohen et al. 2007). Darvill (1995) identified the role of heritage sites and projects in improving wellbeing, linked to their ability to generate a sense of belonging and to connect people to the present by using the past. This research has provided evidence of how these places act as mechanisms for personal and communal interaction with one another and with their environment (Tacon and Baker 2019, Forestry Commission 2013, Cohen et al. 2007, Darvill 1995).

Qualitative research into the social impacts of community heritage projects, such as those of archaeological sites, heritage sites, historic sites, and landscapes provides a starting point for understanding the potential of heritage to impact on wellbeing (Jones 2004, Bender 1998, Nissinaho and Soininen 2014, Darvill et al. 2018, 2019, Power and Smyth 2016, Waterloo Uncovered 2019, Everill et al. 2020, Kiddey and Schofield 2011, Finnegan 2016, Belford 2012, Simpson 2010, Sayer and Sayer 2016). These research projects encompassed a broader definition of heritage including both the tangible and intangible, which enabled a deconstruction of heritage values to be considered in relation to public feelings and attitudes towards 'their' heritage (J. Thomas 2004). Bender's work at Stonehenge, Jones's research at the Hilton of Cadboll, and Nissinaho and Soininen's of Finland's 'Adapt a Monuments Scheme' provided initial evidence of the impact of heritage sites on wellbeing and mental health (Jones 2005, Bender 1998, Nissinaho and Soininen 2014). Sayer's (née Simpson) research into the social value of community archaeology excavations in United Kingdom and United States highlighted key traits including building new relationships, supporting opportunities and desire to learn and developing confidence (Simpson 2010). This research identified the potential of heritage projects to negatively impact on wellbeing, including social relationships, and increasing negative emotions related to stress and anxiety (Simpson 2010).

Research into the social values of heritage places has indicated that many of the values are associated with that of positive wellbeing, including:

- Enhanced sense of community.
- Developing self-esteem and confidence.
- Collective participation.
- Psychological benefit of ownership and dedication to the project.
- Understanding lifecycles.
- Provide new knowledge to support conservation and maintenance.
- Provide meeting sites to discuss personal matters.
- Community empowering spaces.
- Supporting community relationships and breaking down barriers.
- Reintegration of disparate groups back into society.
- Reduction of anti-social behaviour.

The majority of recent research into the impact of heritage sites and landscapes on personal wellbeing and happiness has been done as part of wider surveys relating to the social value of heritage (National Heritage Lottery Fund 2015, Historic England 2017, Historic England 2022, Reily et al. 2018, National Trust 2018, Rosemberg et al. 2011). In the United Kingdom, these surveys have been undertaken by large national government organizations, including Historic England, English Heritage, National Trust, and government bodies such as the Department for Culture, Media, and Sport (Reily et al. 2018, Fluck 2015). These surveys suggest key trends in wellbeing associated values from visiting the historic environment that include: bringing people together; acting as a meeting place for broader community; providing a place of interaction to support socialization and create friendships; providing a place to support people to learn about culture, diversity, and history; and supporting people to understand themselves and empathize with others (Reily et al. 2018, Historic England 2015, 2017, DCMS 2017, Fluck 2015, What Works Centre for Wellbeing 2017, Lakey et al. 2017, National Trust 2018, National Heritage Lottery Fund 2015, National Heritage Lottery Fund 2018, Historic England 2022). Interaction with heritage sites and landscapes can help overcome barriers and support the regeneration of communities, improving belonging, pride, and skills (Pennington et al. 2018, Darvill 2019).

Individual case studies relating to specific heritage projects and sites and their impact on wellbeing have been less researched than that associated with museums. Quantitative wellbeing surveys of heritage places include Chi, Cai, and Li (2017), Sayer (2015, 2018), Heaslip (2019), Sofaer et al. (2021), and Brizi et al. (2023). Chi, Cai, and Li's research into the impact of a World Heritage Site in Lijiang China on residents' wellbeing identified different impacts on wellbeing in specific populations, which included the negative impacts of tourism on the wellbeing of the indigenous community (2017). It indicated that residents who reported high levels of economic satisfaction, sense of community, and satisfaction with social environment had higher levels of wellbeing than those who did not (Chi, Cai, and Li 2017). Sayer's research into the impact of community archaeological excavations on individual and community wellbeing linked the impacts of archaeology to the NEF five ways to wellbeing (Sayer 2015, Aked et al. 2008). It highlighted specific and measurable positive and negative changes to participants subjective wellbeing after participant in heritage projects, including 'forming connections to people and the world around them' and 'increasing personal learning' (Sayer 2015). It also indicated that involvement in heritage projects could negatively impact individual subjective wellbeing, which included increased anxiety, hostility, and distress (Sayer 2015, 2018).

Sayer's subsequent research into the impacts of specific archaeological related activities on school aged children indicated trends in archaeology's impact on wellbeing, including the positive impact that activities that included hands on practical learning and supported ownership had on wellbeing (Sayer 2018). It highlighted that archaeological activities that were controlled and lacked freedom of choice could have a negative impact on wellbeing and personal happiness (Sayer 2018). Heaslip's research into the impact of the Human Henge project on participants wellbeing, indicated the project measurability improved participants wellbeing and mental health (Heaslip 2019). Results suggested participant felt more positive about themselves, more relaxed, and closer and more connected to people (Heaslip 2019: 123). Sofaer's et al. large-scale survey into the impacts of visiting UK heritage sites after Covid on individuals eudaimonic and hedonic wellbeing identified key themes including heritage sites' ability to support physical and psychological capabilities, social connections, ontological security, and trust (2022).

Specific localized qualitative heritage wellbeing surveys have included those carried out by Galeotti (2016), Power and Smyth (2016), Wheatley and Bickerton (2017), Ripanti (2022), Finneran and Welch (2022), Johnston (2019), Clifton (2019), and Darvill (2019). Galeotti's research of the cultural heritage landscapes in Italy indicated that heritage had positive impacts on wellbeing through improving community cohesion, social inclusion, and equality (Galeotti 2016). Power and Smyth's research into the UK NHLF-funded community-led heritage landscape project indicated the positive and negative impacts of these projects on wellbeing. Positives included: bringing people together, engagement, discovery, personal learning, new skills, belonging, delight, curiosity, passion. Negative impacts included frustration, anxiety, patience, group tensions, disappointment, and personal demands and culture clashes (Power and Smyth 2016). Research into heritage landscapes impact on individuals wellbeing often identifies how they have supported meaning and place making by evoking feelings, emotions, and self-reflection, and contributed to peoples' sense of security, belonging, identity, and self-esteem (Clifton 2019, Darvill 2019, Finneran and Welch 2022, Ripanti 2022).

Recently, many quantitative and qualitative research projects focused on the impact specific archaeological excavations and initiatives have on veterans' wellbeing and mental health and their ability to support rehabilitation (Finnegan 2016, Nimenko and Simpson 2014, Dobat et al. 2019, Everill et al. 2020, Waterloo Uncovered 2019). Nimenko and Simpson's and Dobat et al.'s qualitative evaluations and Finnegan and Everill et al.'s quantitative evaluations have all provided evidence that participation in excavation had positive impacts on veterans' social values associated with wellbeing and noted marked improvements in the wellbeing and mental health of participants (Nimenko and Simpson 2014, Finnegan 2016, Everill et al. 2020, Dobat et al. 2019). Everill et al.'s research noted archaeological project's ability to reduce negative feelings, increasing positive feelings, supporting learning, and build confidence and pride (Everill et al. 2020: 223).

Dobat et al.'s qualitative research of metal detecting veterans with disclosed and undisclosed mental health problems indicated that metal detecting had positive and lasting impacts on mental health and wellbeing and specifically alleviated mental disorders (Dobat et al. 2019: 1). This highlighted change was linked to the mental, sensory, physical, and social nature of metal detecting and associated with five organizing categories: (i) peace of mind, (ii) outdoor life and exercise, (iii) silent companionship, (iv) the thrill of history, (v) self-esteem (Dobat et al. 2019: 5).

These small-scale research projects into veterans' mental health, wellbeing, and archaeology have been used to justify the funding of international, national, and regional veteran-associated archaeological projects including: Waterloo Uncovered, Soldier On, and Operation Nightingale (Walshe, Osgood, and Brown 2012, Waterloo Uncovered 2019, Everill et al. 2020).

Gallou's overview of heritage wellbeing research indicated eight pathways to wellbeing outcomes linked to heritage places' ability to: (i) support identity, (ii) provide an experience, (iii) support healthy behaviours, (iv) reduce anti-social behaviour and support socializing, (v) reduce anxiety and increase positive feelings, and (vi) shape capabilities (Gallou 2022). Research to date has indicated the impact of heritage places on wellbeing and has identified key characteristics for positive wellbeing that can be bolstered and created by these visits, including:

- Personal learning.
- Bringing people together.
- Community cohesion.
- Developing new skills.
- Sense of belonging.
- Pride of place.
- Engagement.

Organizations have developed projects at heritage sites, which directly seek to improve community and visitor wellbeing; examples include

Compton Verney and Human Henge (Darvill 2019, Sayer 2018, Darvill et al. 2019, Price and Keynes 2020). These projects have sought to use heritage sites as a backdrop to support mental health; for example, the 'Human Henge' project is a partnership between public health, heritage organizations, and universities supported by National Heritage Lottery Fund, which used the historical landscapes at Stonehenge to support positive mental health (Human Henge 2018, Darvill 2019, Darvill et al. 2019). Compton Verney Art Gallery and Landscape, supported by the National Heritage Lottery Fund, has used its historic buildings and landscapes to support wellbeing and positive mental health, creating a Dementia Café, Woodland Walks, Heritage Trials, Forest Schools, and Tai Chi in the Landscape (Sayer 2018, Desmarais et al. 2018). These projects have adapted and applied public health perspectives and collaborated with a team of multi-disciplinary professionals to tackle mental health issues and provided research to support the professional to understand the role heritage can play in improving wellbeing (Chatterjee and Camic 2015, Darvill 2019). Comprehensive analysis of heritage sites and specific wellbeing projects is a work in progress, but early anecdotal commentary from volunteers, visitors, facilitators, and staff suggests that visitors undertaking specific activities at these heritage sites have experience positive changes to their mental health and wellbeing (Desmarais et al. 2018, Pennington et al. 2018, Darvill et al. 2019).

Research into the impact of heritage sites on wellbeing has begun to indicate what makes heritage unique: for example, the DCMS Taking Part Survey suggested that involvement in historical and heritage sites had higher rates of life satisfaction than involvement in sports (Fujiwara et al. 2014, DCMS 2016, Pennington et al. 2018, Gallou 2022, Historic England 2022). These surveys have suggested that high levels of visiting, interest, attachment, and investment in heritage sites has resulted in over 75.8% of the United Kingdom's population visiting UK heritage sites in 2017, and high membership rates of heritage organizations and sites, including 5 million members of National Trust and 1 million members of English Heritage (DCMS 2017, Historic England 2017, National Trust 2018). This indicates a positive association by the public with heritage and the potential national impact of heritage on wellbeing. The National Trust's psychological survey of brain activities compared historical places and objects and concluded that 'places rather than objects with a story created greater emotional response' (National Trust 2018).

Increasingly, results from heritage wellbeing research are being linked to and corresponded with New Economic Foundation five ways to wellbeing: supporting connections to the world around, forming social relationships, feelings of belonging, ability to self-reflect, work towards common good, and constructing identity (Sayer 2015, Soafer et al. 2021, Dobat et al. 2019: 15). Yet research also indicates that the public associate specific wellbeing values to heritage sites, including creating connections to places, creating a sense of belonging, supporting traditions, and creating personal and family memories (Sayer 2015, Soafer et al. 2021, Dobat et al. 2019, Darvill et al. 2018, Brizi et al. 2023). Public engagement with heritage place and with their specific activities and public associations have unique impacts on wellbeing values, which are different from other heritage assets based on specialist activities and associations, including belonging, tradition, and memory, which specifically impact on individuals' wellbeing and differs from other heritage assets.

Site Analysis

Six case studies from United States, United Kingdom, and India formed the basis of quantitative and qualitative research. These heritage sites were Chatsworth House (UK), Priory Gardens (UK), Biltmore Estate (US), Nemours Estate (US), Bangalore Fort (India), and Tipu Sultan's Palace (India). The following section provides the results of individual case studies and serves to illustrate contextual and localized nuances in changes to specific visitor wellbeing indicators resulting from specific experiences at museums and art galleries

Chatsworth House and Estate

Chatsworth Estate, Derbyshire (UK), encompasses a Grade 1 listed house, gardens, landscape, and farm. The civil parish with a total area of 5.2 km includes smaller villages of Edensor, Beeley, and Pilsley in the Conversative parliamentary constituency of the Derbyshire Dales. The combined demographics for Beeley and Chatsworth Parish indicate a population of 195, with 48% males and 52% females; 54% were over 45; and 81% were economically active (ONS 2021). The house was first built in the 1552 by Sir William Cavendish, as a result of his wife Elizabeth Talbot's wish to live closer to her home county. Of this original building only the hunting lodge survives, which is believed to have been built in 1580 (Chatsworth House Trust 2023). Over the centuries there have been various additions to, and rebuilding of, the house and landscaping of gardens by Capability Brown (Chatsworth House Trust 2023). The house was first opened to the public by the 11th Duke and Duchess of Devonshire (1920–2004) (Chatsworth House Trust 2023). The house currently belongs to the Devonshire family and since 1981 it has been owned and managed as part of Chatsworth Trust, a charitable foundation that aims to ensure the preservation of the house, its essential contents, and its gardens for the general benefit of the public (Chatsworth House Trust 2023).

The cost of visiting the Chatsworth Estate varies depending on which areas of the estate the visitor wishes to access: adult access to the house, garden, farmyard, and playground costs £29, house and gardens £26, gardens and farmyard and playground £18, garden £15, and farmyard and playground £7. Visitors can engage in the heritage, history, and art at Chatsworth through self-guided tours, guided tours, and audio tours, and through engagement with staff who located in each room.

Analysis

117 visitors to Chatsworth completed the online survey before and after visiting. The demographic data from participants, indicated that over 86% were females and the majority, over 49% were over 50; 90% did not consider themselves to have a disability. 83% described themselves as white and most participants (80%) were visiting with family. Chatsworth attracts on average of over 500,000 visitors a year; 79% did not consider themselves to have a disability or long-term illness, 94% described themselves as white, 78% indicated they were females and 25% males, 22% were aged 45–54, 29% were aged 55–64, and 49% were over the age of 65. Demographic comparison indicated a slightly higher percentage of survey participants were female, did not have a disability and were white; age figures were roughly comparable.

Quantitative Analysis

PANAS Negative Affects mean scores indicated a reduction in all negative PANAS associated emotions and an overall reduction of −9.85% in visitors PANAS negative mean average score after visiting Chatsworth (Appendix, Fig. 7). After visiting Chatsworth, visitors had statistically significant decreases and reduction in variability in feeling jittery ($p<0.003$), guilty ($p<0.012$), scared ($p<0.037$), nervous ($p<0.005$), distressed ($p<0.00$), and irritable by ($p<0.000$) (Appendix, Fig. 9). There was a statistically significant decrease but increase in variability in feeling afraid ($p<0.045$) (Appendix, Fig. 9). Results suggested that visitor experiences at Chatsworth positively, significantly, and measurably impacted personal wellbeing in the component of emotional wellbeing, by supporting a reduction in negative feelings.

PANAS positive affect visitor mean scores indicated average increase by 2.21% across all measures and increase in mean score in eight of the measures after visiting Chatsworth (Appendix, Fig. 8). After visiting, visitors experienced a statistically significant increase and reduction in variability in feeling active ($p<0.006$) (Appendix, Fig. 9). Results indicated that visitor experiences at Chatsworth overall supported emotional wellbeing, with increased positive feelings and improved vitality of visitors.

The Modified Wellbeing Scale indicated a positive increase in visitors mean average scores of 0.60% and an increase in nine mean visitor scores after visiting Chatsworth (Appendix, Fig. 10). After visiting Chatsworth, visitors experienced a statistically significant decrease and reduction in variability in feeling life was worthwhile ($p<0.007$) (Appendix, Fig. 11). Indicating visitors' experiences at Chatsworth had a negative change to personal wellbeing in the component self-esteem.

Visiting Chatsworth had slight negative impacts although not significant to visitor wellbeing specifically linked to the component of personal wellbeing and subcomponent of positive functioning (autonomy, meaning and purpose, and engagement) and emotional wellbeing (levels of personal happiness).

PANAS positive and negative, and MWS results indicated that people visiting Chatsworth felt a measurable and significant improvement and change in personal wellbeing as a direct result of their experiences. Visitors indicated Chatsworth had a positive impact on emotional wellbeing, reducing negative feeling, and improving positive feelings. Visitors also indicated a significant improvement to vitality, as they felt more active after visiting Chatsworth. Visitors to Chatsworth experienced significant negative impacts to self-esteem and resilience feeling life was a less worthwhile, positive functioning, feeling less attentive after their visit. Chatsworth supported visitors to be active.

Qualitative Analysis

The free commentary space filled in by visitors to Chatsworth indicated common and recurring themes in their perception of the most meaningful experience at Chatsworth, including: being outside (in nature, fresh air, gardens, views, space, physical exercise) (n = 51); spending time with family and friends (n = 32); culture (seeing house, history, art, and theatre) (n = 30); Chatsworth staff and volunteers (n = 17); peace, tranquillity, and atmosphere (n = 13). Free comment space recurring themes included: enjoyment of visit (n = 24); garden/nature/outside space (n = 13); culture (house/history/art) (n = 10), positive comments about staff (n = 9); reduction of stress/anxiety, relaxing (n = 5); beauty (n = 4), and visitor key words emotive words included enjoying, beautiful, lovely, special, and friendly.

Thematic analysis of visitor commentary based on NEF five ways to wellbeing indicated that visiting Chatsworth impacted participants in the specific areas, which included to connect, to take notice, and be active. Visitor comments indicated Chatsworth provide a space and through various family engagement activities supported families spend quality time and connect with each other, it was commented that 'Chatsworth is part of the fabric of my life/childhood it means a lot to our family' (AV). Visitors frequently noted how the gardens and house supported them to take notice of the world around them, how they felt inspired by visiting the house and gardens, and one visitor commented that taking notice of the garden provided them with 'inspiration and our sheer joy of nature!' (AV). Visitors commented that visiting Chatsworth, specifically spending time in the grounds and gardens, had provided them with the space and motivation to be active and this had been 'physically good' (AV) and was 'energizing' (AV).

Thematic analysis of free comment space indicated that visitors subjectively perceived visiting Chatsworth to have impacted their personal wellbeing and social wellbeing. In relation to personal wellbeing, a visitor commented 'I undoubtedly experienced feelings of enhanced wellbeing as a direct result of going to Chatsworth' (AV). Visitors' comments indicated that improvement to personal wellbeing was linked to four specific components: resilience and self-esteem, emotional wellbeing, vitality, and positive functioning. In the component of resilience and self-esteem, participants noted that visiting Chatsworth supported them to forget their worries and troubles; for example, after family bereavement and chronic pain conditions a visitor stated, 'Chatsworth allowed me to focus away from that for a few precious hours, I'll never forget this day' (AV). Other visitors commented it 'allowed me to leave behind my troubles for a while' (AV) and it supported their mental health 'this visit was very well timed for me having very recently been diagnosed with anxiety. It was a tonic' (AV). Participants felt emotional wellbeing had been improved as they described their experiences as 'multisensory', 'beautiful' 'inspiring', 'calming', and 'relaxing' (AV). The gardens were frequently referenced, with one visitor commented 'the gardens made me feel good, relaxed and happy' (AV). Visiting was stated as being 'energizing' and 'physically good for our wellbeing though tiring for all of us!' (AV), which is linked to improved vitality. In relation to positive functioning, visitors' comments reflected how Chatsworth had supported positive functioning, including meaning and purpose as they described the experience as being 'a worthwhile visit' (AV) and autonomy as they noted the 'freedom to enjoy' (AV).

Visitor commentary reflected an improvement in social wellbeing in the components of trust and belonging and supportive relationships due to visiting Chatsworth. Visitor experienced 'friendly' staff and volunteers at Chatsworth that enabled them to develop supportive relationships, feel respected, supported, and treated with kindness, 'Incredibly kind and actually it meant a huge amount to me and my mum to see people genuinely caring thoughtful and compassionate . . . . it made me feel very proud to be close to a place like Chatsworth' (AV) and 'everyone we interacted with were lovely. Really friendly' (AV). Visitors noted they felt a wider sense of belonging and trust, specifically belonging to Chatsworth and commented on the positive impact their experiences had on their sociocultural identity and sense of belonging, for example on visitors noted after visiting they 'felt proud to be British' (AV).

Visitors noted the varied heritage practices at Chatsworth and a holistic approach to heritage engagement that utilized the landscape, nature, art, architecture, and history to provide a diverse heritage experience that positively impacted their wellbeing '[the]combination of experiencing the delightful decorations in the house, enjoying a lovely walk in the gardens (and being outside in nature always makes me feel good) and spending that time with my mum and enjoying the experience together' (AV).

The visitor comments indicated negative impacts of their experiences at heritage sites linked to personal wellbeing, specifically emotional wellbeing, and positive functioning; these included high costs for tickets, the number of people, and crowding and could be linked to the statistically significant quantitative decrease in feeling life was worthwhile and feeling attentive after visiting Chatsworth.

Qualitative results indicated that visiting Chatsworth improved personal and social wellbeing 'I undoubtedly experienced feelings of enhanced wellbeing as a direct result of going to Chatsworth' (AV). Visitors indicated their social wellbeing had been improved, and Chatsworth had provided a social space to develop supportive relationships through place to spend time and quality interactions with family and friends. Visitors felt supported to develop trust and belonging through relationships developed with family, and the respect and sense of belonging that was engendered between staff, volunteers, and visitors. Chatsworth supported visitors' personal wellbeing, including vitality in providing a space to be physically active and reduce stress, anxiety; emotional wellbeing, high frequencies of positive feelings, including enjoyment, and feeling peace and tranquillity; positive functioning, and visitors felt absorbed in engaging in visiting culture and heritage. Chatsworth negative impacts on visitors' wellbeing were associated with accessibility issues, lack of choice, requirement for more detailed information about history and story of the house, and a lack of educational material about the house. Qualitative results indicated that Chatsworth supported visitors to be active, to connect, and take notice of world around them.

Summary

Triangulation of quantitative and qualitative results indicated that Chatsworth positively impacted visitors' personal and social wellbeing, and measurable positive changes to individual subjective wellbeing can be directly linked to the NAW indicators. After visiting Chatsworth, there was a measurable improvement to visitors' social wellbeing, including in the components of belonging and trust with people and places, and supportive relationships with family and friends. After visiting Chatsworth, there was a measurable improvement in personal wellbeing including in the components of emotional wellbeing, reducing negative feeling and increasing positive feeling; resilience and self-esteem, supporting visitors to feel they can overcome grief and heal from loss; positive functioning, improved through supporting visitors to feel more engaged and to have autonomy; and vitality, significantly improved by providing an outdoor space and facilities for visitors to be physically active. Yet it is also worth noting that results also indicated measurable and significant impacts of personal wellbeing in the component of self-esteem and resilience and positive functioning. Results indicated that Chatsworth supported visitor wellbeing through providing experiences that supported the NEF five ways to wellbeing: it encouraged visitors to be active, to connect with, and to take notice of the world around them.

Priory Gardens

Priory Gardens (PG) is a Scheduled Ancient Monument located in the historic centre of Dunstable; a town located in the county of Bedfordshire

(UK). Dunstable currently has a population of nearly 40,000 residents and is in the conservative constituency of South West Bedfordshire (ONS 2021). Priory Gardens is located on a prehistoric and Roman trade route. In 1100, Henry I re-founded the town of Dunstable and founded an Augustinian Priory on the Priory Gardens site and built a Royal palace opposite the site (Dunstable Town Council 2018). After the dissolution of the monasteries, all that remained of the Priory was the Priory House: the other buildings and gardens were levelled. Between the seventeenth and nineteenth centuries Priory House and Gardens had a number of different owners (Dunstable Town Council 2018). In 1839, Robert Munt and Thomas Brown brought the house and land, they retained Priory House, and created the formal gardens which are visible today (Dunstable Town Council 2018). In 1947, the local authority compulsorily purchased Priory House and Gardens as a public space, and the park and gardens were opened to the public on 24 May 1947 (Dunstable Town Council 2018). Since the purchase of Priory Gardens by Dunstable Town Council there has been a continuous programme of improvement, and archaeological excavations have been carried out on the site (Dunstable Town Council 2018).

Priory Gardens is a local community asset that is free for visitors to enter and is open to the public every day and all year around. Next to Priory Gardens are the listed buildings of Priory Church and Priory House. The Gardens are funded and managed by Dunstable Town Council, and maintenance is carried out by a range of volunteers and through working with a number of organizations, including Historic England. Dunstable and Priory Gardens are both supported by Historic England's High Street Heritage Action Zone programme.

Quantitative Analysis

Seventy visitors took part in the in-person before and after surveys at Priory Gardens. The demographics were as follows: 53% of respondents indicated described their gender as female and 47% as male, 62% of respondents were under the age of 50, 90% did not consider themselves to have a disability, 90% described themselves as white, and 53% were visiting with family. Priory Gardens does not collect demographic or visitor data, so no comparative data is available.

PANAS negative affect visitors' mean scores increased on average across all measures by 2.2% and six mean scores increased after visiting Priory Gardens (Appendix, Fig. 12). There were no statistically significant changes to specific negative measures (Appendix, Fig. 14). Results suggested that visitors experienced a measurable but not significant increase in negative feelings, linked to a negative impact on personal 'emotional' wellbeing after visiting Priory Gardens.

PANAS positive affect visitors' mean scores increased across all measures by an average of 4.38% and increased for seven measures after visiting Priory Gardens (Appendix, Fig. 13). After visiting Priory Gardens, visitors experienced a statistically significant increase and reduced variability in feeling attentive ($p<0.043$) and determined ($p<0.019$) (Appendix, Fig. 14). Results indicated that Priory Gardens improved the personal wellbeing of visitors, specifically in the categories of positive functioning linked to supporting engagement, emotional wellbeing linked to an increase in positive feelings, and self-esteem and reliance supporting an increase in feeling determined.

MWS visitors' mean scores increased by an average of 4.71% across all measures and improved in fourteen measures after visiting PG (Appendix, Fig. 15). After visiting PG, visitors experienced statistically significant increases and reduction in variability (SD) for satisfied with my life ($p<0.013$), sense of accomplishment ($p<0.013$) and connected to people around me ($p<0.039$), and a statistically significant increase with no change in variability for happy with life ($p<0.006$) (Appendix, Fig. 16). Results indicated that visitors to PG experienced measurable and significant improvements to positive functioning, satisfying life, and social wellbeing in the indicator of supportive relationships.

After visiting PG, visitors' mean scores for PANAS negative and positive and MWS indicated a slight but not statistically significant negative impact to their personal wellbeing specifically linked to the component of emotional wellbeing (Appendix, Figs 14 and 16). Visitors felt more jittery, guilty, hostile, scared, upset, irritable after visiting PG and indicated a reduction in feeling life was worthwhile linked to self-esteem and resilience (Appendix, Figs 14 and 16).

Quantitative analysis indicated that visitors to PG felt a measurable improvement and change in wellbeing in the indicators of personal and social wellbeing as a direct result of their experiences. Visiting PG significantly and measurably improved the personal wellbeing of participants, specifically in the components of emotional wellbeing, supporting visitors to feel less negative feeling and more positive and increased feeling of happiness; positive functioning, with visitors feeling more attentive; life satisfaction, with visitors feeing a sense of satisfaction with lives as a result of visiting PG; self-esteem and resilience, supporting visitors to feel more determined and an increased sense of accomplishment. Visits to PG measurably and significantly improved participants' social wellbeing, specifically in facilitating the development of supportive relationships with family and friends, with visitors feeling more connected to people around them. Quantitative results indicate that PG measurably and significantly supported participants to take notice in the world around them, to feel more attentive and to connect to people in their local community.

Qualitative Analysis

Thematic analysis of visitor commentary indicated recurring themes in most meaningful experience, which included: friends/family time/children (n = 23), improved wellbeing (n = 20), enjoyment/fun (n = 10), other people (n = 11), activities (n = 12), community/local area (n = 9), learning (n = 6), day out (n = 8), history/archaeology (n = 3), free (n = 3). Key emotive words used by visitor in free commentary space included seeing, love, child, enjoy, family and people.

Thematic analysis of visitor commentary indicated that PG provided experiences that supported visitors in the NEF five ways to wellbeing, to connect, to take notice, to be active, and to keep learning. Participants commented that visiting Priory Gardens had supported them to connect with each other, through providing space and place to 'spend time with family' (AV) and friends, to 'come together to interact with each other' (AV) and 'meet up with friends for a shared day out' (AV). The 'free activities' (AV) and doing 'activities with my family' (AV) and that they 'enjoy[ed]' (AV) supported them to connect through spending time together. PG provided space and a place to 'get outside' (AV) and supported visitors to 'feel more connected to the local area' (AV) and to feel 'very blessed by where I live' (AV). Visiting PG supported 'people to come together in a community way' (AV), 'talk to people' (AV), in what was described as 'an invigorating, community-spirited way' (AV); comments indicated that PG supported engagement between staff volunteers and visitors and encouraged people to develop new relationships.

Commentary reflected how PG had supported visitors to take notice, enabling them to 'enjoy' (AV) taking part in 'new experiences' (AV), 'do lots of interesting things' (AV) and see 'interesting displays' (AV). It supported visitors to take notice of the 'local area' (AV) and people around them, supporting them to 'chat to different people' (AV). Visitors commented that visiting PG had supported them to keep learning, with comments which noted they had 'learn[t] new things' (AV). Participants indicated that the interactive activities and diverse displays enabled them to 'relate to history' (AV) and how 'it made me look at the Gardens and Priory and see what was there and the history of the place. If I hadn't done this with my daughter, I don't think I would have known anything about the links to Henry 8th or gone into the Priory' (AV). Visitors commented that learning had been supported through staff sharing 'knowledge and experience with others' (AV) and being provided with learning activities that were 'fun' (AV). Commentary indicated how coming to PG had supported them to be active, through encouraging them to 'get out and do something different' (AV), be 'out in the fresh air' (AV), and 'spend time outside' (AV).

Analysis of visitor commentary indicated that visiting PG impacted personal and social wellbeing indicators. The personal wellbeing of visitors was improved in the components of emotional wellbeing, as visitors frequently noted 'enjoyment' (AV) and that visiting was a 'lovely' (AV) day and day out. Enjoyment was linked to the 'different activities' (AV) on offer at Priory Gardens, having a 'family day out' (AV) and 'shared day out together' (AV), 'seeing people engaging normally' (AV), and 'interacting with each other' (AV). Improvement to participants emotional wellbeing was noted by numerous participants who commented the 'visit had a positive effect on me' (AV) and they felt 'more cheery' (AV) and it was 'very lifting' (AV). These positive feelings were linked to spending time together and engaging in fun activities which provide the backdrop for a 'happy event' (AV) and an 'enjoyable family day out which impact[ed] my wellbeing positively' (AV); positive functioning, with visitors noting that 'experiences' (AV) at PG positively influenced their levels of engagement with each other and with the history and heritage of place. They indicated that engagement had been supported by 'getting outside and do[ing] activities with family and others' (AV), 'getting out and doing something different' (AV), and 'seeing different displays for the medieval period' (AV). Improvement in positive functioning was associated with PG providing a meaningful and purposeful day out, 'teaching children' (AV) and supporting children to 'learn' (AV) about history; regarding vitality, visitors indicated that Priory Gardens had supported them to go 'outside' (AV) and to 'spend time outside' (AV) in the fresh air, and one visitor noted that after being physically active they were 'tired' (AV); and self-esteem and reliance: one participant noted the impact that visiting PG had on their confidence about their future career choices 'it has confirmed my choice of career' (AV).

Visitor commentary indicated visiting Priory Gardens impacted social wellbeing in the components of trust and belonging as visitors noted Priory Gardens supported them to feel a sense of belonging to Dunstable and the local community: 'it's confirmed and boosted my sense of being very blessed by where I live' (AV), 'I feel very blessed to live in their friendly, supportive, and interesting community with such a great mix of cultures' (AV), and '[I] feel more connected to [the] local area' (AV); supportive relationships, with visitors commenting that Priory Gardens had provided a supportive space and a place to 'talk to people' (AV), to 'chat[ting] to different people' (AV), specifically 'local people' (AV), and the space to 'spend time with family' (AV) and see 'a good number of friends' (AV). Visiting PG supported families, friends, and community members to build supportive relationships by providing space and a place to spend time together during 'fun' (AV) experiences.

Qualitative results indicated PG supported participants take notice, to connect, be active, and to keep learning and improved participant personal and social wellbeing. Social wellbeing was improved in the components of supportive relationships, providing a social space and time to develop existing and new supportive relationships and interact with family, friends, and members of wider community; trust and belonging, supporting a sense of belonging to a community and to a place. Personal wellbeing was improved in the components of emotional wellbeing as participants noted high frequencies of positive emotions, including feeling happy; positive functioning, participants felt experiences at PG had meaning and purpose, supporting them to learn new things and new skills; engaged with the history and heritage; self-esteem and resilience, supporting a participant to make choices about their future; and vitality, participants indicated by getting outdoors they felt more active.

Summary

Triangulation of qualitative and quantitative results indicated that Priory Gardens had a measurable and significant positive impact on visitor subjective wellbeing and positively impacted the NAW indicators. Visits to Priory Gardens measurably and significantly improved visitors' personal wellbeing, including visitors' emotional wellbeing, positive functioning, satisfying life; and measurably improved visitors self-esteem and reliance, and vitality. Visits to Priory Gardens measurably and significantly improved visitors' social wellbeing, in the component supportive wellbeing and measurably improved trust and belonging. Yet it is also worth noting results also indicated measurable negative impacts of personal wellbeing in the component of self-esteem and resilience and emotional wellbeing. Priory Gardens supported the NEF five ways to wellbeing, supporting participants to be active, connect, keep learning, and to take notice of the world around them.

Biltmore Estate

Biltmore Estate is a historic estate located in Ashville, United States. It is situated in Buncombe Country in the state of North Carolina, which has Republican senators and a Democrat governor; Ashville has a population of approximately 69,000 people (United States Census Bureau 2020). Biltmore Estate is the largest historic estate in the United States and is owned by the Biltmore Company, a private company set up by the descendants of the Vanderbilts and Cecils, the original owners. The estate is managed as a commercial enterprise and is open to the public.

In 1887–1888, George Vanderbilt bought the land in which Biltmore was built (Biltmore Company 2023). Construction of Biltmore 'Chateau', a 250-room house started in 1889, and the landscaped grounds and gardens were designed by landscape architect Frederick Law Olmsted. The house was officially opened in 1895 by George Vanderbilt, and in 1924 his daughter Cornelia Vanderbilt married John Francis Amherst Cecil. In 1930, Cornelia and John Cecil opened Biltmore House to the public to support tourism and generate income to preserve the estate during the Depression (Biltmore Company 2023). Biltmore house stored art during World War II including work from the National Gallery of Art in Washington, DC (Biltmore Company 2023). In 1960, William A. V. Cecil, Sr, joined his brother to support the self-sufficiency and preservation of the estate, which included planting vines for wine, and by 1983 Biltmore Estate Wine Company had been established (Biltmore Company 2023). In 2001, the Inn on Biltmore Estate was opened, followed by the Antler Hill Village in 2010 and in 2015 the Village Hotel (Biltmore Company 2023).

The estate is open to the public seven days of the week with differing seasonal operating hours. Ticket prices are managed dynamically by season and day of the week and vary depending on the type of visit (audio-guided house tour, expert-guided house tour, self-guided tour of the gardens, and entry to the grounds) and which areas of the estate visitors wish to access (house, gardens, grounds, or special exhibition). Admission is divided into just the House $89 (May–September), $76 (October to November); Audio Guided ($89–119), Audio Guided Plus (with Monet and Friends Experience) ($119–149), Expert Guided Visit Plus ($354–384), and just Gardens and Grounds $55–85. Ticket prices include access to Antler Hill Village and winery including a free wine tasting, and parking. Biltmore House offers all guests audio tours as part of the ticket price; this is the only mechanism by which most guests are provided with narratives and interpretations relating to the site. Except for titles of key rooms and additional material within rooms for specific artwork, including Monet and Renoir, there is limited interpretative material available and few interpretative text panels located in the house or gardens.

Quantitative Analysis

724 visitors completed online surveys both before and after visiting. Demographic data from Biltmore Estate indicated that 79% of respondents described their gender as female, 67% were over +50, 96% did not consider themselves to have a disability, and 94% described themselves as white. Of these visitors 75% were visiting with family, 99% were not annual pass holders, 98% were not guests at Biltmore Hotel, 65% made regular visits to cultural heritage sites, and 63% were first-time visitors. Biltmore does not collect demographic data from visitors so comparative analysis with survey participants was not possible.

PANAS Negative Affects visitors' mean score indicated an average decrease of –2.31% across all measures and eight scores decreased (Appendix, Fig. 17). Visitor experiences at Biltmore resulted in statistically significant decreases and reduction in variability in feeling irritable ($p<0.00$), distressed ($p<0.00$), and upset ($p<0.010$) (Appendix, Fig. 19). These results suggest that Biltmore positively impacted visitors' emotional wellbeing by reducing negative feelings.

PANAS Positive Affects indicated visitors' mean score increased by 0.30% across all measures and increased for nine measures after visiting Biltmore (Appendix, Fig. 18). Visitor experiences at Biltmore statistically significantly increased and reduced variability (SD) in visitors feeling inspired ($p<0.047$) (Appendix, Fig. 19). Results indicated an improvement to positive functioning after visiting Biltmore and a slight improvement to visitors' positive feelings and improvement in emotional wellbeing.

MWS results indicated visitors' mean average scores decreased by –0.01% across all measures, and on four measures mean scores increased after visiting Biltmore Estate (Appendix, Fig. 20). Visiting Biltmore resulted in statistically significant increased and reduced variability in visitors feeling able to connect to people around me ($p<0.013$), and significant increased and increased variability in visitors feeling able to trust people ($p<0.000$) and supported by each other ($p<0.000$) (Appendix, Fig. 21). Visiting Biltmore resulted in a significant decrease and reduction in variability in visitors feeling able to have freedom of choice ($p<0.015$), and a significant decrease in visitors feeling able to overcome challenges ($p<0.047$) and having the opportunity to learn new things ($p<0.004$) (Appendix, Fig. 21). These results indicated Biltmore had a positive impact of visitors' social wellbeing, specifically linked to supportive relationships and trust and belonging, yet had a negative impact on visitor personal wellbeing, in the components of positive functioning (meaning and purpose, autonomy) and self-esteem and resilience.

Visitor experience at Biltmore negatively impacted visitor personal wellbeing in the subcomponent of positive functioning, specifically *autonomy* with visitors feeling they had less freedom of choice and meaning and purpose, with people feeling they had less opportunity to learn after their visit. After visiting Biltmore, visitors felt lower levels of self-esteem and resilience, specifically less able to overcome challenges.

Quantitative results indicated that Biltmore measurability and significantly supported visitor to connect and take notice. Results indicate that people visiting Biltmore felt a measurable improvement and change in personal and social wellbeing as a direct result of their experiences. There was measurable improvement to visitors' social wellbeing as visitors felt they had developed a sense of belonging and trust with each other and had developed supportive relationships with friends, family, and staff at Biltmore. Personal wellbeing was improved as visitors experienced measurable improvement to positive functioning and felt more inspired and engaged by Biltmore's history and landscape after visiting, and there was an improvement to emotional wellbeing; visitors noted feeling less negative after the visit.

Qualitative Analysis

Qualitative results indicated recurring themes in visitors perceptions of what the most meaningful about their visit, which included; the house (n = 255); the site's history (n = 151); the grounds/gardens/greenery/plants (n = 105); spending time w family/friends (n = 94); the Vanderbilt family (n = 72); learning/education (n = 61); enjoyment (n = 61); seeing architecture/design/craftmanship (n = 46); beauty (n = 44); viewing the conservatory (n = 42); Christmas (n = 35); art (n = 33); the views/scenery (n = 27); viewing servants' quarters (n = 22); audio tour/guide (n = 21); interactions with staff (n = 21); viewing the library (n = 16); commitments to sustainability/restoration/preservation (n = 16); sense of awe/wonder/appreciation (n = 16); Vanderbilts' staff (n = 14); sense of peace/relaxing/calm (n = 10); interesting (n = 10).

Thematic analysis of visitor free commentary space indicated key recurring themes, which included enjoyment (n = 56); suggestions for improvements (n = 48); the house (n = 28); audio guides/tours (n = 28); the grounds/gardens (n = 27); staff (n = 25); beauty (n = 24); the Vanderbilt's life/life-style (n = 23); returning visitor, or will return in future (n = 22); history (n = 18); education/learning/knowledge (n = 15); thanks (n = 16); crowd size / lines (n = 14); and COVID measures (n = 11). Key emotive words in commentary included enjoyed, inspiring, beautiful, meaningful, wonder, and kind.

Qualitative analysis of comments indicated that visitors felt Biltmore had impacted wellbeing in the NEF five ways to wellbeing indicators of connect, take notice, and keep learning (Aked et al. 2008). Visitors indicated that engagement at Biltmore supported them to connect to people working on the estate in both the past and present, commenting 'It made me think a lot about lifestyles and class privileges back in the 1800s and early 1900s. I feel like life was so hard for the laborers who built the estate, employees that managed the transportation of the residents and their many visitors who had to use trains, horses and buggies, and the staff responsible for running the estate people' (AV). Visitors commented that spending time in the gardens and estates had 'connected [them] to nature' (AV). Visitors indicated that their experiences at Biltmore had supported them to take notice, enabling them to feel 'energized, interested and engaged during our entire day's visit' (AV). Visitors stated they 'loved' and 'enjoyed' learning about the history of the estate and Vanderbilt family and 'learning about the past [brought] perspective to [the] present' (AV) and that they 'loved learning how much George Vanderbilt travelled in his life' (AV), indicating that engagement with and experiences at Biltmore has supported participants in the indicator to keep learning.

Biltmore impacted visitors' personal wellbeing and social wellbeing indicators. Visitors indicated that an improvement to emotional wellbeing was linked to spending time in garden and grounds, house, and winery. Visitors frequently stated the gardens and grounds were 'beautiful', 'inspiring', 'calming', and 'relaxing', and improvement to emotional wellbeing was linked to spending time in these outdoor locations 'I loved the vibe of the grounds! Calming and relaxing' (AV), and 'my time there was very peaceful and beautiful' (AV). The winery was described by visitors as a place to 'relax and enjoy' (AV) and the house was a place of 'happiness' (AV) and 'it filled my eyes and heart with joy' (AV). One visitor commented the value post-Covid of the experience at Biltmore that 'I realized how much I missed having an experience like I had yesterday' (AV). Visitors indicated that Biltmore supported their reliance and self-esteem: for one visitor it acted as a healing mechanism after bereavement 'Ave Maria played once again. It was very sweet this time and I was glad to think of my mother once again' (AV). Visitors' comments reflected that Biltmore had improved their positive functioning, and they noted high levels of engagement indicating '[Biltmore] took me back in time' (AV), that it was 'an inspiring home' (AV) and a 'magical place' (AV). Visitors indicated development of meaning and purpose, one stating that 'the conservatory 'inspired me to be more creative with my gardening' (AV) and it was a 'wonderful and memorable visit . . . it was worth the time and expense to see such a beautiful piece of American history' (AV). One visitor noted that there were 'so many options for things to do so everyone would have something they could enjoy' (AV), with a positive impact to autonomy.

Biltmore impacted visitors' social wellbeing, one visitor comment indicated that staff and guides at the site were 'friendly' 'helpful', and 'nice', and treated visitors with kindness, 'everyone that we interacted with across the estate (employees) were incredibly kind and helpful, made you feel welcome' (AV) and 'the hospitality of all the people in the estate was unbelievable, they were lovely. Really friendly' (AV). Visitors indicated a sense of belonging and trust to Biltmore and its history. Visiting had positive impacts on their sense of cultural and national identity and belonging: 'Biltmore is a beautiful piece of Americana' (AV). Biltmore provided visitors with a place to spend quality time and shared experiences with family, to help build supportive relationships. Visitors commented that 'this visit brought my family together so close' and the experience was 'awesome to share with the family' and supported creating positive family memories: 'being at the estate was something we will all remember for a very long time' (AV).

Visitors' commentary indicated their experiences had negative impacts on their personal wellbeing; this included high costs, number of people, specifically crowding (linked to Covid and safety measures including social distancing and facemasks), accessibility to some areas of house 'there is too much that is off limits to guests' (AV), choice and requirement for more detailed information about history, including on the audio tour and signs, the story of the house, and limited tickets for the winery.

Qualitative results indicated that Biltmore supported visitors to take notice, to connect and to learn. Visitor experiences at Biltmore impacted personal and social wellbeing. Biltmore positively improved social wellbeing, developing a sense of belonging and trust through the respect and sense of belonging that was engendered between staff, guides, and visitors, and Biltmore provided the space, time, and the creation of shared memories to develop support relationships with family and friends. Visitor experiences at Biltmore improved participants personal wellbeing, including emotional wellbeing, with visitors relating their experiences with positive feelings, including being happy, relaxed, and calm. Biltmore improved visitors' positive functioning, supporting them to learn, and enabling visitors to feel a sense of accomplishment, engagement, and that they had a meaningful and purposeful visit. Biltmore developed visitors' resilience and self-esteem in supporting them to develop strategies to cope with grief.

Summary

The triangulation of qualitative and quantitative results indicated that Biltmore had a positive impact on visitors' subjective personal and social wellbeing. Visiting and engaging with Biltmore positively impacted the NAW indicators: visitor experiences at Biltmore measurably and significantly improved personal and social wellbeing. Biltmore had most measurable impact on improving visitors' social wellbeing, specifically in developing a sense of belonging and trust, and in facilitating the development of support relationships with family and friends. Experiences at Biltmore positively impacted, and measurably and significantly improved visitors' personal wellbeing, specifically emotional wellbeing, positive functioning, and self-esteem and reliance. Biltmore supported subjective wellbeing through providing experiences linked to the NEF five ways to wellbeing, enabling visitors to connect and take notice and keep learning with the world around them (Aked et al. 2008).

Nemours Estate

Nemours Estate is located in Montchanin near Wilmington, within the county of New Castle, in the State of Delaware. The Democratic Party controls the office of governor and both chambers of the state legislature of Delaware. New Castle County has a population of 570,000. The demographics of the county are 65% white, African American 26%, 10% Hispanic or Latino, 6% Asian, 0.5% American Indian/Alaska Native, 2.6% two or more races, 0.1% native Hawaiian or other pacific islander, 52% Female, 16% over 65 (United States Census Bureau 2020). The Nemours historic estate is part of the Nemours Foundation, a children's hospital, and is funded by the foundation (Nemours Foundation 2023). It is not required to make a profit and does not take paying guests as part of its financial structure. The estate, which was built by Alfred I. Dupont for his second wife Alicia in 1907, is a seventy-seven-room mansion, chauffeurs' garage, and 200-acre gardens and grounds (Nemours Foundation 2023). The house initially sat on a 3,000-acre plot, of which approximately 2,000 acres remains (Nemours Foundation 2023). The mansion was built in late eighteenth-century French style, and the estate is named after the French town that Alfred's great-great-grandfather represented in the French Estates General (Nemours Foundation 2023). The house was opened to public in 1977 (Nemours Foundation 2023).

Admission is free for hospital patients, families, and staff. For other visitors, entry costs $18 and at present there is no café or gift shop. Visitors engage in the estate through either a self-guided or guided tour of gardens and/or house. Staff are located in each room and interact with visitors; they provide visitors with non-scripted key facts, short histories, stories about paintings, objects, and families and answer questions about any objects. Each of the rooms in the house contains descriptive interpretation panels, which discuss the use of the room, a short human story and historic photographs, and some objects have smaller factual interpretative panels and object searches for children. There is no set route for exploring upstairs, downstairs, and the ground floor; many of the rooms only have key items roped off, allowing visitors free movement around the rooms.

Quantitative Analysis

Forty-seven visitors completed in-person before and after surveys. 72% of respondents described their gender as female, 59% were over the age of 60, 91% did not consider themselves to have a disability, 87% described themselves as white, and 51% were visiting with family. Nemours Estate does not collect demographic information of visitors to enable comparison of the survey sample with all visitors; in 2021–2022 it received a total of 3,058 visitors.

PANAS Negative Affects mean visitors' scores indicated a reduction in across all measures by an average of –13.81% and nine measures decreased after visiting Nemours (Appendix, Fig. 22). Visiting Nemours statistically significantly decreased and reduced variability in visitors feeling guilty ($p<0.012$) and nervous ($p<0.049$) and decreased feeling jittery ($p<0.000$), distressed ($p<0.000$), and irritable ($p<0.001$) (Appendix, Fig. 24). PANAS negative results indicated a reduction in negative feelings linked to improvement in emotional wellbeing.

PANAS positive affect visitors' mean scores increased on average by −13.81% and all mean scores increased after visiting Nemours (Appendix, Fig. 23). After visiting Nemours, visitors experienced a statistically significant increase and reduction in variability in feeling strong ($p<0.013$), inspired ($p<0.041$), and attentive ($p<0.049$) (Appendix, Fig. 24). PANAS positive affect results indicated Nemours Estate supported visitors' emotional wellbeing in the categories of positive functioning (meaning and purpose, and engagement) and self-esteem and reliance (strong).

Modified Wellbeing Scale visit mean scores increased across all measure by an average of 1.89% (Appendix, Fig. 25). After visiting Nemours, visitors experienced an increased in twelve MWS measures and decrease in variability (SD) for ten measures, yet none of these changes was statistically significant (Appendix, Fig. 26).

Quantitative results indicated that visiting Nemours Estate resulted in a measurable and significant improvement to visitors' personal wellbeing. Quantitative improvement to personal wellbeing was in the components of emotional wellbeing, as Nemours supported visitors to feel less negative about their lives, including significantly less guilty, nervous, jittery, distressed, and irritable; positive functioning, with visitors feeling more inspired and attentive because of visiting Nemours Estate. Visitors felt stronger which could be linked to self-esteem and reliance. Visiting Nemours Estate measurably and significantly supported visitors to take notice in the world around them.

Qualitative Analysis

Thematic analysis indicated recurring themes in visitors' commentary regarding their most meaningful experience during their visit, these included: grounds/views/nature ($n = 26$); history ($n = 11$); learning ($n = 9$); house ($n = 7$); family/friends ($n = 6$); staff/volunteers ($n = 3$). Analysis also revealed recurring themes/answers to free comment: space ($n = 20$); out ($n = 2$); beauty ($n = 4$); gardens/grounds ($n = 5$); history ($n = 3$); friends and family ($n = 3$); enjoyment ($n = 5$); staff ($n = 2$). Thematic analysis of key emotive words from visitor commentary found that the most frequently used words were: garden, beauty, enjoy, family, grounds, nature, friend, and walking.

Qualitative analysis of visitor comments indicated visiting Nemours Estate impacted NEF five ways to wellbeing as follows: visitors indicated that their experiences at Nemours estate had supported them to connect with each other, providing the space and place to spend time with their family and friends, to 'laugh[ing] with my friends' (AV), 'outdoor walks with friends' (AV), 'meeting old friends' (AV), 'conversing with friends' (AV), and to 'sitting with family looking at the gardens' (AV). Visiting Nemours provided visitors with a focal point for shared experiences for 'sitting with family looking at the gardens' (AV) and a space to trigger family memories 'the music room reminded me of my grandmother's . . . lots of happy memories from the past' (AV) and to 'come back share our experience with our family and friends' (AV). It supported visitors to 'connect with history' (AV) through 'taking a visit in the past' (AV). The 'additional information offered by room attendants in the home' (AV) and 'hearing the family stories' (AV) helped visitors 'envision the life of the people who lived here' (AV). Nemours estate supported the development of connections between staff and visitors, 'terrific staff' (AV), including through 'talking to survey volunteers' (AV).

Visitor commentary indicated the visit supported people to take notice, and to feel 'absorbed' (AV) by 'sitting on [a] bench absorbing the beauty' (AV) during their visit to the gardens, grounds, and house. Visitors appreciated the role of Nemours Estate, specifically the grounds and gardens, in providing a space to appreciate, enjoy, and be 'close to nature' (AV) and 'looking at nature and enjoying all the surprises that wildlife gives' (AV). Through taking notice and being in nature the visitors felt the experience was 'very relaxing' (AV) and 'meaningful' (AV). Seeing the interior of the house enabled visitors to notice the history of the site and the 'elegance and the intricacies of the property' (AV).

Commentary indicated visiting Nemours Estate supported people to keep learning, specifically about the people and place to learn 'about the DuPont's intelligence' (AV), 'about the gardens' (AV) and 'history of the place' (AV). Visitors commented that their experience at Nemours had 'increased their knowledge of the DuPont Family' (AV) and being told hidden and more diverse narrative was appreciated, including 'hearing about the servants' good treatment' (AV) and 'the explanation[s] of how each room was used' (AV). This had for some 'renewed [their interest] in history' (AV). Visitors commented that being outside in the grounds and gardens and 'actively walking through the beautiful environment' (AV) and 'walking through the grounds' (AV) supported them to be active.

Visitors noted that Nemours was a great place for 'running' (AV) and a 'beautiful [place that] children can explore' (AV).

Visitors' commentary indicated that Nemours had impacted their personal wellbeing and social wellbeing. In relation to personal wellbeing, visitors frequently mentioned their 'enjoyment', and visitors' comments indicated specific impacts to wellbeing based on relaxation and calm 'I found myself enjoying the beauty but finding little quieter and simpler corners in which to reconnect with nature and relax' (AV) and 'feeling how the water calmed me down' (AV). Enjoyment was linked to 'enjoying history and art' (AV) and the 'beauty' (AV) and 'beautiful gardens' of Nemours Estate. Emotional wellbeing was supported by the 'beautiful scenery' and by being 'close to nature' and how visitors 'loved the visit to the estate' (AV).

Visitor comments reflected how engagement with Nemours Estate supported positive functioning. Positive functioning of visitors was supported through the presentation of history and stories of the house, grounds, and gardens, which enabled visitors to 'appreciate the work for this place' (AV) and acquired 'an appreciation of how this family used their engineering skills to create amazing things in the late 1800 early 1900' (AV). Comments indicated that Nemours was a meaningful and purposeful place for visitors, in which visitors could learn about the 'history of the place, the DuPonts and the gardens' (AV). Visitors' comments reflected that coming to Nemours Estate supported their vitality, including to 'walk' and 'run' (AV).

Analysis of visitor commentary suggested that Nemours impacted social wellbeing in the components of trust and belonging. Visitors commented they felt a sense of 'connect[ion] with history' (AV) and related this to 'hear[ing] the family stories' (AV). Visitors noted that the visit supported them to connect with people in the past through 'Realiz[ing] that people long ago did almost the same things that we do today . . . . Working together to accomplish one goal' (AV). Visitors indicated visiting Nemours helped them develop supportive relationships, as it provided a place to 'meet old friend[s]' (AV) and spend time with loved ones. Visitors noted the role of the 'terrific' (AV) and 'great' (AV) volunteers and staff who provided information throughout the visit and helped them to feel personally connected to the history of Nemours.

Visitors' comments reflected some negative impacts of Nemours on their personal wellbeing, these related to issues with the tour bus with one visitor commenting they 'felt badly that [the] shuttle tour [was] not better managed. Late arrivals didn't count, people got on first. Some who were there first didn't get on' (AV) and they 'need [a] system to not let too many people on bus. [They need a] system to enforce first come first [serve] on the bus (AV). Other negative impacts to personal wellbeing related to the noise of machinery used to manage the gardens and indicated the 'need to use battery operate weed walkers to reduce nerve wracking noise' (AV).

Qualitative data indicated that visiting Nemours improved visitors social and personal wellbeing. Nemours supported social wellbeing; it provided a social space and time to develop supportive relationships and quality interactions with family, friends and loved ones. Visits supported trust and belonging, through enabling relationships and memories to be developed with loved ones, and by the respect and sense of belonging engendered between staff, volunteers, and visitors. Visitor experiences at Nemours estate supported personal wellbeing in the indicators: emotional wellbeing, as visitors indicating high frequencies of positive emotions, including feeling happy, relaxed, and calm after visiting; positive functioning, visitors felt inspired and engaged with the history of Nemours, they felt the visit had meaning and purpose, and they felt encouraged and inspired to learn. Visitors noted that their active and outdoors visit had improved their vitality. Qualitative results indicated that Nemours Estate supported visitors to take notice, connect, be active, and to keep learning.

Summary

Triangulation of qualitative and quantitative results indicated that Nemours had positively impacted visitors' personal and social wellbeing in the NAW indicators: Nemours had measurably and significantly improved visitors' social wellbeing, specifically in developing a sense of belonging and trust, and in facilitating the development of support relationships with family and friends. Experiences at Nemours measurably and significantly improved visitors' personal wellbeing, specifically emotional wellbeing, positive functioning, self-esteem, and resilience and vitality. Nemours supported subjective wellbeing through providing experiences linked to the NEF five ways to wellbeing, enabling visitors to be to connect, be active, take notice, and keep learning with the world around them.

Bangalore Fort

Bangalore Fort is in Bengaluru, in the Karnataka district (India). The site is owned and managed by Archaeological Survey India, which is a government agency. Bangalore is a city located in the southern Indian state of Karnataka, with a population of over 10 million, and currently governed by the Bhartiya Janata Party (Government of Karnataka 2023). Bangalore Fort was built as a mud fort as in 1537 by Kempe Gowda I, the founder of Bangalore, as part of the Vijayanagar Empire (UNESCO 2014). In 1761, Hyder Ali replaced the mud fort with a stone building and his son Tipu Sultan added further improvements in the late eighteenth century (UNESCO 2014). It was damaged during the Anglo-Mysore war in 1791. On 21 March 1791, the army of the British East India Company, led by Lord Cornwallis, captured the fort in the siege of Bangalore during the Third Mysore War (1790–1792). At the time the fort was a stronghold for Tipu Sultan (UNESCO 2014). In 1811, the British converted the fort into quarters for troops and by 1831 the British had assumed direct administration of Mysore, and most of the buildings that were part of the fort were converted into public offices (UNESCO 2014). The fort became the main ammunition supply store for British troops located outside Bangalore (UNESCO 2014). In October 1888, the garrison was vacated, and the Bangalore Fort was handed over to the civil administration authorities (UNESCO 2014). The city has grown around the fort, currently the fort's Delhi gate and two bastions are the primary remains. There is a marble plaque that commemorates the place where the British breached the fort's wall and captured it.

The site is open to the public seven days a week, from 10 am to 5 pm. Entry to Bangalore Fort is INR 15 (approximately 15 pence) for local or India resident visitors, and INR 200 (approximately £2) for foreign or non-resident visitors. Engagement in Bangalore Fort is either through self-guided tours or tour guides. Interpretative features on the site includes interpretation panels, with simple information about architectural features and historical facts.

Quantitative Analysis

Eighty-two visitors completed before and after in-person surveys. The majority 59% of respondents described their gender male, 81% of respondents were aged 20–29, 100% did not consider themselves to have a disability. There was no visitor demographic data available for Bangalore Fort overall to compare with survey participant demographics.

PANAS negative visitors' scores indicated an average increase across all measures of 1.65% for all negative emotions and a decrease in five scores, yet there was no statistically significant changes to any scores after visiting BF (Appendix, Figs 27 and 29). Results suggested visiting Bangalore Fort increased visitors' negative feelings and had a negative impact on visitors' emotional wellbeing.

PANAS positive affect visitors' mean scores increased across all measures by an average of +7.81% and mean scores for all measures increased after visiting Bangalore Fort (Appendix, Fig. 28). After visiting Bangalore Fort, visitors experienced a statically significant increase and reduced variability (SD) in feeling attentive ($p<0.001$) and excited ($p<0.002$) (Appendix, Fig. 29). There was a statistically significant increase with no change in variability for determined ($p<0.001$), and statically significant increase and increased variability enthusiastic ($p<0046$) inspired ($p<0.006$) and interested ($p<0.007$) (Appendix, Fig. 29). Results indicated that visiting Bangalore Fort significantly and measurably improved visitors' personal wellbeing, in the indicators of emotional wellbeing through supporting positive feelings and emotions; and positive functioning, specifically through improving levels of engagement.

Modified Wellbeing Scale visitor mean scores increased on average across all measures by +4.45% and all measures mean scores increased after visiting Bangalore Fort (Appendix, Fig. 30). Visitors' experiences at Bangalore Fort significantly increased and reduced variability in feeling happy with my life ($p<0.002$), able to overcome challenges ($p<0.018$), and the opportunity to learn new things ($p<0.002$) (Appendix, Fig. 31). Visitors experienced a significant increase and an increase in variability for belong to a neighbourhood ($p<0.004$), interested in the world around me ($p<0.008$), connected to the people around me ($p<0.012$), and satisfied with my life ($p<0.045$) (Appendix, Fig. 31). Results indicated that visits to Bangalore Fort significantly and measurably improved the personal wellbeing of visitors, including to their emotional wellbeing, positive functioning (meaning and purpose) and self-esteem and resilience, and impacted life satisfaction and social wellbeing, in the components of belong and trust and supportive relationships.

Visiting Bangalore Fort significantly and measurably improved visitor personal wellbeing and social wellbeing. The analysis indicated a measurable and significant improvement to visitors' emotional wellbeing, and they felt more positive and happier with life; positive functioning, with visitors feeling more attentive, interested, inspired, excited, attentive, and enthusiastic after visiting; invoking self-esteem and resilience, visitors felt measurably and statistically more determined and able to overcome challenges. Visitors experienced significant improvement in social wellbeing specifically in facilitating the development of supportive relationships with family and friends and trust and belonging, as visitors experienced improved feelings of belonging to a neighbourhood. Quantitative results indicate Bangalore Fort supported visitors to take notice, to keep learning, and to connect to people around them.

Qualitative Analysis

Thematic analysis of visitor commentary indicated recurring themes in most meaningful experiences: good/excellent/enjoyment n = 24; survey n = 6; photographs n = 4; disappointed n = 10; beauty n = 3; price n = 2; small n = 6; history n = 10; family/friends n = 2. Word count analysis indicated that place (27), good (19), visit (19), fort (9), history (9), much (9), see (8), expect (7), nice (7), Bangalore (6), palace (6), small (6), and time (5) were the words used with highest frequency. Key emotive words in visitors' free commentary space were: enjoy, beauty, friends, family, good, and nice.

Thematic analysis of visitor commentary indicated that BF supported visitors to connect with each other, providing them with a place to spend time with friends 'is nice to visit with friends' (AV) and 'it was a great time in Bangalore Fort with friends' (AV). Visits to Bangalore Fort enabled visitors to take notice of a site that was described as a 'little gem' (AV), and take notice of the fort's history, architecture, and art 'marvellous architecture' (AV) and 'a must visit place for history lovers and people who love ancient paintings. The palace is well maintained, and the architecture and design blows your mind' (AV). Visitors commented it was the 'beautiful' (AV) and 'nice environment and preservation' (AV) and 'art forms' (AV). Experiences at Bangalore Fort supported visitors to 'feel inspired' (AV) and 'enjoyed' their visit, and that they 'enjoy[ed] every moment with happiness' (AV). Visitors commented that engaging with Bangalore Fort had supported them to keep learning and to 'learn new things' (AV) about the history and architecture of the fort, commenting visiting 'helped [them] to know the ancient architecture and the materials used for construction at that time' (AV) and 'this is the best site to learn about history' (AV). Visitors noted it was a good place to 'explore the history of this place' (AV). And 'explore about unknown history of this fort' (AV).

Commentary indicated Bangalore Fort impacts visitors' personal wellbeing and social wellbeing. In relation to personal wellbeing, visitors frequently noted in free space its impact on emotional wellbeing and improvement of positive feelings, for example '[I] enjoyed every moment with happiness' (AV) and that they had a 'great time' (AV). Visitors associated improvement of positive emotions with their 'good experience' (AV) of the physical space and the 'beautiful environment' (AV) and 'nice environment' (AV), for example 'nice place', 'beautiful', and 'good place' (AV). Visitors' experiences at Bangalore Fort supported their positive functioning, specifically through engagement with its 'history' and 'architecture' (AV), which was supported by the presentation and preservation of the fort, with comments including 'the old-fashioned way of life and the status of the city under the control of several monarchs. As soon as you enter the location, you will discover a wealth of information' (AV) and 'see[ing] the old fort still existing' (AV). Visitors commented they were 'inspired' (AV) to 'explore' (AV) the fort's history, which provided a meaningful and purposeful day to 'learning new things' (AV) and to 'learn about history' (AV). Visitors' scores indicated a positive impact on their social wellbeing, specifically in developing supportive relationships through providing a place to connect with friends 'it was a great time in Bangalore Fort with friends' (AV) and 'it is nice to visit with friends' (AV).

Qualitative data indicated that visiting Bangalore Fort improved people personal and social wellbeing. It improved visitors' social wellbeing, providing a social space and time to develop existing and new supportive relationships and quality interactions with friends. It improved personal wellbeing; emotional wellbeing with high frequencies of positive emotions, including feeling enjoyment; positive functioning, as visitors felt engaged with the history and architecture of the fort and that the place had meaning and purpose. Qualitative results indicate that the Bangalore fort supported visitors to take notice, connect, and keep learning.

Summary

Triangulation of qualitative and quantitative results from visitors to Bangalore Fort indicated that the site had a measurable and significant positive impact on NAW indicators, including social and personal wellbeing. It significantly improved visitors personal wellbeing, specifically in the components of emotional wellbeing, positive functioning, and self-esteem and resilience. Visiting Bangalore Fort supported social wellbeing, specifically in facilitating the development of supportive relationships with friends and wider community and a sense of belonging and trust in their local neighbourhood. Visitor experiences supported the NEF five ways to wellbeing and visitors were helped to connect, keep learning, and to take notice of the world around them.

Tipu Sultan's Summer Palace

Tipu Sultan's Summer Palace (TSSP) is located in Bangalore, in Karnataka (India), in the urban and busy Krishna Rajendra City market area. Bangalore is a city located in the southern Indian state of Karnataka, with a population of over 10 million, and currently governed by the Bhartiya Janata Party (Government of Karnataka 2023). The Palace is maintained by the Government of Karnataka and is a popular tourist attraction in the city (Government of Karnataka 2023). The foundations of Tipu Sultan's Summer Palace were built by Hyder Ali, the Sultan and ruler of the kingdom of Mysore (Government of Karnataka 2023). Following Hyder Ali's death in 1782, the construction of the building, a two-storey structure built from teak, and gardens was completed in 1791 by his son, Tipu Sultan, and became his family's summer palace. After Tipu Sultan's death, the British administration took ownership and used it as a secretariat (Government of Karnataka 2023). The architecture and artist designs on walls and carved in wood represents the Indo-Islamic tradition and there are walls covered in paintings depicting the grand throne of the Sultan. The Palace was opened to the public in 2005.

The cost of entry to TSSP varies from 15 INR for residents (approximately 15 pence) to 100 INR (approximately £1) for foreign or non-resident visitors. It is open 8:30–17:30 every day. Visitors engage with the site through self-guided tours or guided tours. Located around the site are interpretative panels, with historical information and details about specific areas of the site. On the ground level one room has been converted into a museum, which presents items belonging to Tipu Sultan, including a replica of Tipu's Tiger, which is an eighteenth-century mechanical toy created for Tipu Sultan, and represents a tiger savaging a near life-size European man.

Quantitative Analysis

Ninety-six visitors completed in-person before and after surveys. Demographic data indicated that 51% of respondents described their gender male, 76% of respondents were aged 20–29, and 100% did not consider themselves to have a disability. The site, as part of Archaeological Surveys India, does not collect demographic data so no comparative analysis of the survey demographic data was possible.

PANAS negative affects visitors' mean scores indicated an average increase across all scores of 3.21% and decrease in six of the measures' mean scores after visiting TSSP (Appendix, Fig. 32). None of the changes to PANAS negative emotions mean scores after visiting TSSP was statistically significant (Appendix, Fig. 34). Results suggested that after visiting TSSP visitors felt a measurable improvement in emotional wellbeing and a reduction in negative feelings.

PANAS positive affect visitors' mean scores indicated an average increase of +8.56 and that all positive affect mean scores increased after visiting TSSP (Appendix, Fig. 33). After visiting TSSP, visitors experienced significant improvements and reductions in variability in feeling excited ($p<0.050$), active ($p<0.005$), enthusiastic ($p<0.002$), and significant improvement and increase in variability in feeling strong ($p<0.000$), interested ($p<0.000$), determined ($p<0.004$), attentive ($p<0.005$), inspired ($p<0.007$), and proud ($p<0.015$) (Appendix, Fig. 34). Results indicating visiting TSSP had a significant impact on visitors' personal wellbeing, specifically emotional wellbeing, improving positive emotions; positive functioning, linked to engagement; and in vitality, linked to feeling more active. Results suggested that visitors felt some improvement to self-esteem and reliance (strong and proud).

MWS mean visitor scores increased by an average of 6.15% after visiting TSSP, and mean scores increased across all measures (Appendix, Fig. 35). Visiting the TSSP resulted in a statistically significant increase and decreased variability in visitors feeling they belonged to a neighbourhood ($p<0.000$), connected to the people around me ($p<0.033$), my life is worthwhile ($p<0.003$), and indicated a statistically significant increase (but increased variability in: sense of accomplishment ($p<0.001$), interested in the world around me ($p<0.002$), physically active ($p<0.003$), optimistic about my future ($p<0.017$), I have the opportunity to learn new things ($p<0.018$), satisfied with my life ($p<0.041$), and I can overcome challenges ($p<0.049$) (Appendix, Fig. 36). MWS indicated that visiting the TSSP improved visitors' social wellbeing, specifically belonging and supportive relationships, and personal wellbeing: self-esteem and reliance, and impacted vitality, positive functioning, and satisfying life.

Quantitative results indicated that visits to TSSP significantly and measurably improved personal wellbeing and social wellbeing. It improved personal wellbeing in the components of emotional wellbeing, supporting visitors' positive emotions with a reduction in negative emotions and feelings; positive functioning, with visitors feeling more attentive, interested, inspired, excited, attentive, enthusiastic, and having a greater sense of accomplishment; self-esteem and resilience, with visitors feeling able to overcome challenges; and vitality, with visitors feeling measurably more active because of their experiences at TSSP. Visitors experienced a measurable and significant improvement to their social wellbeing, facilitating the development of supportive relationships with family and friends, as visitors felt more connected to people around them, and trust and belonging, as visitors felt increased belonging to a neighbourhood. Visits to the Sultan's Palace helped visitors to take notice of the world around them, to keep learning, to connect to people around them, and to be active.

Qualitative Analysis

Qualitative analysis of visitor commentary was limited, as few visitors answered the 'most meaningful experience' question. Visitor commentary from TSSP had high levels of negative recurring themes which included: disappointment (13), logistic and entry process (17), and more maintenance and preservation needed (20). Only three of the recurring themes were positively good: excellent and enjoyment (43) and those related to architecture (5) and history (18). Free comment space indicated recurrent words that included: place (41), good (31), visit (18), experience (15), palace (15), maintain (10), proper (10), historic (8), nice (8), time (8), well (8), entry (7), much (7), people (7), preserve (7), expect (6), history (6), like (6), need (6), new (6), ticket (6), environment (5), know (5), online (5), peaceful (5), person (5). Key emotive words in free space commentary included good, experience, nice, and peaceful.

Thematic analysis indicated that visits to TSSP impacted NEF five ways to wellbeing in enabling visitors to connect, take notice, and keep learning. Visitors commented that they felt connected to the 'vibrant history' (AV). The site's presentation of its heritage supported intergenerational connections and connects between staff, tour guides and visitors; commentary indicated this was supported by 'the people here are also welcoming' (AV), and staff supported a 'new generation to know and learn from them' (AV). Visitor experiences indicated that they were helped to take notice of the historic environment and of their 'vibrant past' (AV). Visitors were supported to be 'interactive with a new place' (AV) and to 'feel inspired' (AV) by the 'beauty' (AV). Visitors to TSSP 'enjoyed coming to this place' (AV) and were able 'to imagine that this was done more than 200 years ago! The beauty is astounding' (AV). Visitors to TSSP frequently commented that their experience at the site had helped them to keep learning and their visit had improved their knowledge of history, commenting it had supported 'knowing its history and heritage' (AV) and 'it was nice to know our history and heritage of the place' (AV). Visitor comments indicated a new awareness of TSSP's history had enabled visitors to 'learn about new historic details and about the architecture' (AV), 'learnt also about how thing are in old times' (AV), and 'learnt about new places and their history' (AV).

Visitors commented that experiences at TSSP had impacted their personal wellbeing and social wellbeing. Visitor comments reflected the positive impacts of TSSP on emotional wellbeing, linked increased positive feeling and emotions and that they 'enjoyed coming to this place' (AV), 'good experience' (AV), and 'nice and peaceful' (AV). The positive effect on emotional wellbeing was associated with TSSP providing a place for 'peace', 'relaxation', and 'calm', and being a 'peaceful place to visit' (AV), a 'peaceful environment' (AV), 'relaxing environment'(AV), 'calm environment' (AV). This positive impact on emotional wellbeing was often related to the site's having gardens and outdoor space within a busy urban centre, noting the site's 'beauty' (AV), 'beautiful garden space and the palace', and this provided a 'peaceful place to visit' (AV) and the 'pleasant atmosphere' (AV). Visitors' comments reflected an improvement to their positive functioning after visiting TSSP, specifically that 'good experiences' (AV) and 'new experiences' (AV) had positively supported engagement with the 'history' and 'architecture' (AV). Visitors indicated that engagement with the palace supported them to feel 'inspired by how they have preserved this palace made out of wood and how they have built this structure long back' (AV). Comments indicated the 'preservation' of the site and 'how they have preserved the place from a long time' (AV) supported by engagement, and provided a meaningful and purposeful day out, and a place to learn: 'nice place to interact with the building and learn about new information from the history of the building' (AV) and to 'learn about history' (AV). Visitor comments occasionally indicated visiting TSSP supported their social wellbeing in the components of supportive relationships, by providing a place to connect with staff 'the people here were very welcoming' (AV).

Word count and frequency analysis of visitor questionnaires showed high levels of negative comments, which indicated that there was some negative impact of TSSP on visitors' personal wellbeing. Visitors noted issues with the preservation of the site 'it's really sad that most the intricate carving has not been preserved. The visitors still make marks on the wall'

(AV) and 'looking at the current state, it's not maintained properly' (AV). One visitor related poor preservation to the current national political narrative about religion 'one of the greatest historical places of the entire world but local government and local people not taking care of much because of political and religious issues' (AV). Other negative visitor experience related to entry procedure and interpretive methods, which was described as 'entry is chaotic. No line or queues. Also, some people are not digital sound, they are not able to pay. Also please keep more notices and description about palace' (AV).

Qualitative results indicated that visiting TSSP improved personal and social wellbeing. The site improved social wellbeing, providing a space and place to see people and connect with staff at the palace and help develop supportive relationships. It supported personal wellbeing, around emotional wellbeing, with visitors noting high frequencies of positive emotions, including feeling enjoyment and peace, and positive functioning, with visitors feeling their experience had meaning and purpose, and supported them to engage with the history and architecture of TSSP. Visits to TSSP helped visitors to take notice, connect, and learn.

Summary

Visits to TSSP measurably and significantly impacted the NAW indicators for personal and social wellbeing (Michaelson et al. 2009). Their experience supported visitors in all the indicators of personal wellbeing, including emotional wellbeing, positive functioning, self-esteem and reliance, life satisfaction and vitality, and social wellbeing, specifically supportive relationships. TSSP provided visitor experiences that supported the NEF five ways to wellbeing, including supporting visitors to be active, connect, keep learning, and to take notice of the world around them (Aked et al. 2008).

Comparative Analysis of Heritage Sites

Comparison of visitor quantitative and qualitative data from heritage case studies revealed variation between places, and specific impacts to visitors' subjective wellbeing indicators based on geographic locality (urban/rural), nature of visitor experiences and type of heritage sites (historic estate, archaeological site, or historic building), and types of physical experiences (indoor/outdoor).

PANAS negative affects visitor scores indicated that after visiting four of the six heritage sites there was a measurable reduction to visitors mean average results (Nemours, Chatsworth, TSSP, and Biltmore). Visits to these sites significantly improved visitors' emotional wellbeing through a reduction in negative emotions. The extent of the measurable reduction in negative emotions differed between sites, which indicated the impact of different visitor engagement and experience strategies at heritage sites on emotional wellbeing. For example, visitors to Nemours (−13.81%) and Chatsworth (−9.85%) experienced the largest reductions in mean PANAS negative mean scores and high numbers of significant reductions across the measures, while visitor to TSSP (−3.21%) and Biltmore (−2.31%) experienced smaller reductions in mean scores (Appendix, Fig. 37). At Chatsworth, Nemours, and Biltmore visitors experienced significant reductions to 'feeling irritable' and 'distressed' after their visits. At Chatsworth and Nemours visitors felt significantly less nervous, jittery, and guilty after visiting. Uniquely, at Biltmore visitors felt significantly less upset, and at Chatsworth less afraid and scared after their visit. At Priory Gardens (2.26%) and Bangalore Fort (1.65%) visitors experienced increases in PANAS negative mean scores, yet at these two sites there were no statistically significant change to any PANAS negative measures. These results suggest that visitor experience at Priory Gardens and Bangalore Fort did have a significant impact on negative feelings (Appendix, Fig. 37).

Comparative analysis of quantitative results suggested that only 50% of the sites (all of which were historic estates) had a significant impact on the emotional wellbeing indicator linked to this reduction. Two sites, BF and PG, which were archaeological sites with archaeological remains and landscape features resulted in increased visitor PANAS negative mean scores and no significant changes to any measures. At these sites interpretation was dominated interpretation and presentation heritage to the visitors that focused on leaflets and interpretative panels around key features. At these sites, which did not significantly reduce negative feelings, visitor experiences were centred around outdoor experiences of heritage and did not focus on a singular narrative, but presented broader holistic interpretations of multiple interwoven pasts, and encouraged visitors to explore physical remains independently.

PANAS negative results suggest that historic estates, that provided visitors with a mixture of indoor and outdoor experiences, including landscapes gardens, had the greatest measurable impact on negative emotions. At historic estates interpretation provided a specific story and storied experience of the history of the heritage site centred on the people who lived in the houses and on the estates. Interpretative mechanisms and engagement strategies were more varied and included: recreated and reconstructed rooms; tours, guides and audio guides, books, leaflets, and limited interpretation panels that sought to support visitor interactions with tour guides and volunteers as a mechanism to help visitors engage with heritage. This people-centred approach based on human interactions at historic estates might explain why visitors experience a significant decrease but also increase in some negative emotions, for example increases (although not statistically significant) in feeling irritable, distressed, and nervous.

PANAS positive visitor mean average scores increased at all six heritage case study sites after visiting and suggested a measurable improvement in emotional wellbeing linked to an increase in positive feeling because of visiting heritage sites (Appendix, Fig. 37). TSSP (8.56%) and BF (7.82%) produced the largest increase in PANAS positive mean scores and the highest number of significant changes to specific measures, followed by Nemours (5.45%), Priory Gardens (4.38%), Chatsworth (2.21%) and Biltmore (0.30%) (Appendix, Fig. 37). TSSP had the highest number of statistically significant changes and reduction in variety to PANAS positive visitor scores, including to visitors feeling active, stronger, excited, enthusiastic, attentive, determined, and uniquely to visitors feeling interested after visiting (Appendix, Fig. 34). After visiting BF, visitors experienced statistically significant increases to inspired, attentive, excited, enthusiastic, and determined (Appendix, Fig. 29). On the other hand, visitors to Nemours felt more inspired, attentive, and stronger, and visitors to PG were more attentive, and determined (Appendix, Figs 24 and 14). At Chatsworth the only significant improvement was to 'feeling active' and at Biltmore to 'feeling inspired' (Appendix, Figs 9 and 29).

At the two largest historic estates (Chatsworth and Biltmore) visitors experienced the least positive changes to their positive emotions. At Chatsworth there was a significant reduction in visitors 'feeling attentive' and 'interested' after their visits, linked to a reduction in positive functioning.

PANAS positive results indicated that visitors to all six sites experienced statistically significant improvement to their personal wellbeing in the indicator of emotional wellbeing. Yet there were also differences between historic sites, historic estates, and archaeological sites, for example historic sites (Nemours, Chatsworth, TSSP) were the only locations which supported visitors to experience a significance increase in the personal wellbeing component of vitality and to feel active, possibly as these sites had large grounds, gardens, and outdoor spaces to explore. Archaeological sites (Priory Gardens and Bangalore Fort) were unique in visitors experiencing significant increases in feeling attentive linked to positive functioning (engagement) and 'determined', linked to self-esteem and reliance. Analysis indicates that visitors to Indian heritage sites experienced the most significant positive change to personal wellbeing components of emotional wellbeing, in the indicator of increase in positive emotions and positive functioning in the indicator of engagement.

PANAS results indicated the heritage sites improved visitors' subjective hedonic wellbeing, by positive affecting their positive feelings and reducing negative feelings associated with happiness (Fig. 4.1). They indicated Nemours had the most optimal impact on hedonic wellbeing 'happiness' and had greatest impact on reducing negative feelings and improving positive feelings. Biltmore had least impact on visitors' hedonic wellbeing, and Priory Gardens and Bangalore Fort negatively impacted capabilities associated with visitors' hedonic wellbeing, with negative feelings increasing after visiting. Indian sites of Bangalore Fort and Bangalore fort improved hedonic wellbeing, by increasing positive feelings but not reducing negative feelings, and the large historic estates of Biltmore and Chatsworth had the least impact on improving hedonic wellbeing, linked to improvement in positive feeling (Fig. 4.1).

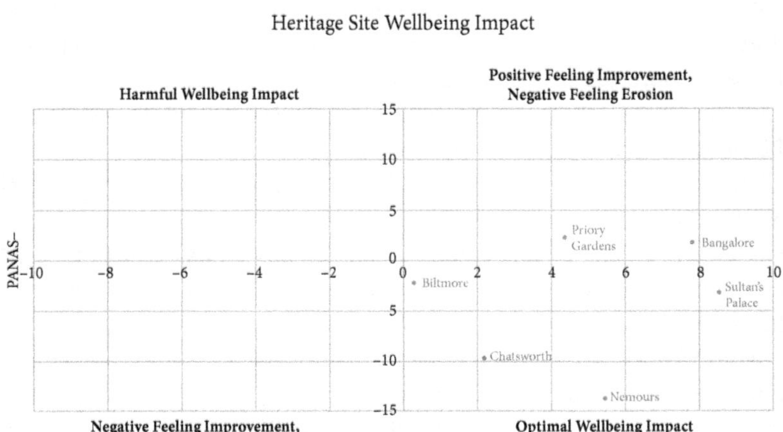

Fig. 4.1 Mapping of heritage and archaeological sites PANAS scores relating to optimal hedonic wellbeing impact.

Comparison of MWS mean visitor results for heritage sites, indicated that except for Biltmore (–0.01%), all heritage site positively impacted and changed visitors overall MWS mean average scores (Appendix, Fig. 37). Visitors to TSSP (5.15%) and BF (4.57%) experienced the greatest positive changes to MWS average mean scores, and the highest number of significant increases to MWS measures after visiting. Visitors to Priory Gardens (4.71%), Nemours Estate (1.89%), and Chatsworth (0.60%) all experienced an increase in mean average scores (Appendix, Fig. 37).

The categorization of heritage and archaeological sites MWS visitor mean average scores based on questions linked to social and personal wellbeing indicators, enabled the comparison between case study sites based on their specific impact on visitors personal and social wellbeing indicators. Results indicated TSSP had the highest impact on visitors' social wellbeing and personal wellbeing, and visitors to TSSP experiences the most optional impact their eudaimonic wellbeing (Fig. 4.2). Whilst Biltmore Estate improved visitors' social wellbeing but resulted in personal wellbeing erosion, and Chatsworth House improved visitors' social wellbeing but had limited impact on their personal wellbeing (Fig. 4.2). Results indicated heritage and archaeological sites improved visitors' social wellbeing, but had less impact on visitors' personal wellbeing, with in some cases, at historic estates negatively impacted visitors' personal wellbeing (Fig. 4.2).

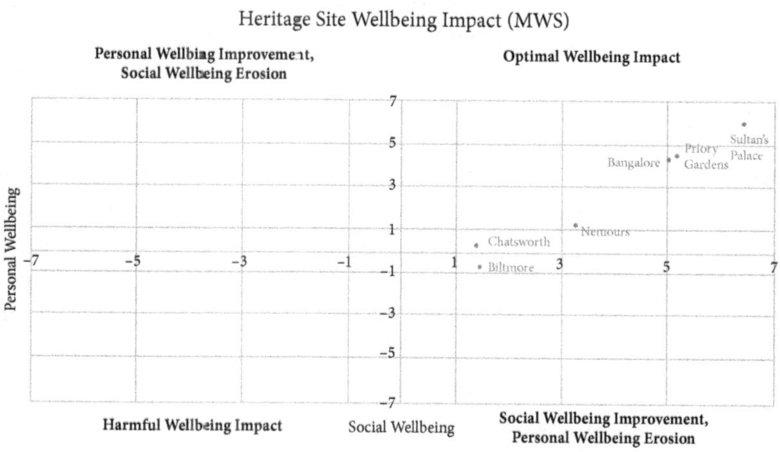

Fig. 4.2 Heritage site wellbeing impact based on MWS social and personal wellbeing indicator scores.

Visitors to TSSP experienced the highest number of statistically significant changes to MWS, including a unique impact on improving visitors feeling physically active and optimistic about the future (Appendix, Fig. 36). Visitors to BF, TSSP, PG, and Biltmore all experienced significant increases in 'feeling connected to people around them' (Appendix, Figs 16, 21, 31, and 36). Visitors to Biltmore, TSSP and PG experienced significant changes to 'feeling a sense of accomplishment' (Appendix, Figs 16, 21, and 36). Visitors to TSSP, BF, and PG experienced significant changes to 'feeling satisfied with life' (Appendix, Figs 16, 31, and 36). Visitors to the archaeological sites of BF and PG experienced significant improvements to 'feeling happy with life' (Appendix, Figs 16 and 31). The only the Indian sites of BF and TSSP significantly increased visitors' feelings 'able to overcome challenges', 'opportunity to learn', 'interested in the world', and 'belonging to a neighbourhood' (Appendix, Figs 31 and 36). Visitors to TSSP and Chatsworth felt 'life was worthwhile' after visiting (Appendix, Figs 11 and 36). There were no significant changes to any MWS categories at Nemours (Appendix, Fig. 26). At Chatsworth the only significant decreased was to 'life is worthwhile', linked to negative impact of self-esteem, and at Biltmore there was a reduction MWS mean score and a negative significant impact on 'freedom of choice' (Appendix, Figs 11 and 21).

MWS results indicated that visitor experiences at historic estates (Chatsworth, Biltmore, and Nemours) resulted in lower overall changes to visitor subjective personal wellbeing, linked to specific hedonic wellbeing capabilities of life satisfaction and eudaimonic wellbeing capabilities of positive functioning and social wellbeing. Archaeological and historical sites, particularly those located in urban areas, smaller site, and those in India (TSSP, BF, and PG) had the most measurable impact on visitors' personal wellbeing on average in relation to MWS categories, including life satisfaction, emotional wellbeing, positive functioning, vitality, self-esteem, and resilience. Different types of heritage sites produced a diversity of personal and social wellbeing impacts; for example, only archaeological sites (Priory Gardens and Bangalore Fort) had significant impacts on personal wellbeing components of emotional wellbeing, specially 'feeling happier with life' which is linked to improvement in hedonic wellbeing.

Quantitative analysis showed that all the heritage sites impacted personal wellbeing. Heritage and archaeological sites in urban areas, particularly in India, had the most significant impact on personal and social wellbeing categories. Visitor scores at TSSP—a site with a garden, standing buildings, a museum and guides showed the most significant and measurable impact on visitor personal and social wellbeing and supported visitors to keep learning, to take notice, to be active, and to connect.

Thematic analysis of visitor commentary indicated that all the case study heritage sites impacted personal wellbeing, and that visits to all sites improved emotional wellbeing and positive functioning. Yet, visitors indicated variation in personal wellbeing indicators affect (vitality, self-esteem and reliance, autonomy) by specific case study heritage sites. These differences seemed to be related to the type and location of the site. Visitor commentary indicated that Chatsworth was the only site where visitors felt they had not been supported to keep learning, which may be linked to visitors to Chatsworth commenting that the information on the interpretative panels was limited, focusing on presenting modern art collection in historic rooms, rather than on architectural and historical narratives of rooms and place. Visitor commentary indicated that only historic estates Chatsworth, Priory Gardens, and Nemours supported visitors to feel active. Comments indicated that large historic estate with indoor and outdoor facilities supported visitors' vitality. Visitor commentary indicated the large historic estates of Chatsworth House and Biltmore Estate supported visitors' self-esteem and resilience, specifically to feel more optimistic, and to consider and feel more confident about their future careers. Visitor commentary at these large historic estates suggested they also supported visitor autonomy linked to freedom or choice and movement around sites. No sites appeared through visitor commentary to support visitors to feel they had a satisfied life or to give.

Visitor commentary indicated that visiting all sites had supported social wellbeing; the indicators impacted varied dependent on the location of each heritage site, specifically its national context. Visitors indicated the impact of all heritage sites on developing supportive relationships between family and friends, although Nemours and Biltmore (US heritage sites) more frequently noted supportive relationships being developed with staff, and tour guides through their interactions with members of staff and shared stories. Visitors to Biltmore, Nemours, PG, and Chatsworth (heritage sites located in the United States and United Kingdom), commented how experiences at these sites supported elements of trust and belonging, specifically to family and national identity. On the other hand, visitors' comments from TSSP or BF did not indicate that they felt a sense of belonging to a place or that trust in people had been impacted on by their experiences.

Qualitative comments indicated that all heritage sites supported visitors to take notice of the world around them and to connect with people and/or places, and with exception of Chatsworth all sites supported visitors to keep learning. Chatsworth House, Priory Gardens, and Nemours Estates supported visitors to be active. Overall qualitative analysis indicated the larger historic estates (Biltmore and Chatsworth) had a wider ranging impact to visitors personal and social wellbeing, and they impacted the broadest variety of wellbeing indicators. Comparative analysis of visitor commentary highted how local context and type of heritage site impact visitor wellbeing. It highlighted potential national differences in perceptions of wellbeing which will be explored in the following chapter.

Conclusion

This chapter has indicated that all heritage sites had a significant and measurable impact on visitors' subjective personal wellbeing and social wellbeing. Comparative analysis of the impact of the case study heritage sites on visitor wellbeing indicated that visiting heritage sites irrespective of location and context had a positive influence on subjective hedonic and eudaimonic wellbeing. Results indicate subtle differences to benefits to wellbeing indicator components and measurable differences between heritage site based on location and nature of site, specifically to components of eudaimonic wellbeing, linked to positive functioning, vitality, and self-esteem and resilience. Archaeological sites and smaller historic buildings and sites located in urban areas had the greatest measurable and significant impact on visitors' subjective hedonic and eudaimonic wellbeing across all indicators. Visitor commentary suggested that visiting heritage sites had the greatest positive impact on visitors' qualitative perceptions of their subjective hedonic wellbeing. All heritage sites supported NEF five ways to wellbeing, yet most evident from this research was the ability of heritage sites to provide experiences that encouraged visitors to take notice and to connect, yet heritage sites in this case study did not appear to support visitors to give, although they had a measurable impact on to keep learning. Only the larger historic estates supported visitors to be active.

Heritage and Wellbeing: The Impact of Heritage Places on Visitors' Wellbeing. Faye Sayer, Oxford University Press.
 DOI: 10.1093/9780191914539.003.0004

5
Museums, Art Galleries, and Wellbeing

This chapter examines how visitor experiences and the nature of presentation and interpretation within museums and art galleries impact visitors' wellbeing, mental health, and happiness. It assesses case studies of museums and art galleries in the United Kingdom and India using the evaluation methodology set out in the previous chapter to provide quantifiable and qualifiable evidence of the measurable impact of visiting this range of museums and art galleries. It identifies through empirical evidence how different museum and art galleries change visitors' personal and social wellbeing and seeks to identify common themes and specific wellbeing indicators that these heritage places effect.

Museums and Galleries Wellbeing Research

Beginning in the 1990s, quantitative museum visitor surveys provided some of the first formal indications of the impact museum visits can have on people's social values, including those later to be associated with wellbeing and mental health. These included large-scale surveys by Merriman and Hooper-Greenhill (Merriman 1991, Hooper-GreenHill 2004, 2007). Merriman's museum survey provided evidence of the values the public associated with visiting museums and identified many factors associated with improving wellbeing, such as personal learning, understanding the present, understanding ourselves, curiosity, interest in the world around us, and appreciation of our lives today (Merriman 1991). Hooper-Greenhill's quantitative survey investigated the educational role of museums and suggested the potential of learning and educational programmes to impact wellbeing, including increased self-confidence and personal learning from taking part (Hooper-Greenhill 2004, 2007). These non-specific museum social value surveys provided tentative glimpses into museums' ability to improve visitor wellbeing (Merriman 1991, Hooper Greenhill 2004, 2007).

Museums and galleries were some of the first heritage organizations to engage in discourse relating to wellbeing, and provided both quantitative and qualitative evidence of their impact on wellbeing and mental health (Thomson et al. 2012, Smiraglia 2015, Chatterjee and Camic 2015, Kindleysides and Biglands 2015). Qualitative research undertaken on specific 'wellbeing' and mental projects at Beamish Museum, Living Museum of the North, Guildford Museum, and the British Museum used conversations, logbooks, open-ended questions, and personal reflections to evaluate the impact that these had on individual wellbeing (Kindleysides and Biglands 2015, Rasbery and Goddard 2011, Phillips 2008). Qualitative evaluation of Holburne Museums Pathways to Wellbeing Programme, a museum-based social prescribing programme indicated improved physical, emotional, and mental wellbeing including improved social skills, better focus and concentration, increased physical activity, feeling valued, feeling connected to other people, feeling a sense of belonging, learning new skills (Ryan 2019, 2022). These projects highlighted that involvement and practical engagement in participatory activities, including object handling and craft activities helped improve wellbeing specifically through supporting personal learning, building friendships, creating social networks, supporting creativity, developing skills and self-expression (Kindleysides and Biglands 2015. Historic England 2017: 42, Phillips 2008, Rasbery and Goddard 2011, Belford 2011, Chatterjee and Camic 2015, Ryan 2019, 2022, Cutler 2022).

The museum sector led the way in implementing formal quantitative public health-based evaluative methods to assess the impact of heritage outreach on wellbeing, which informed future practice (Thomson et al. 2011, 2012, Happy Museum Project 2011, Thomson and Chatterjee 2014, Chatterjee and Camic 2015). For example, researchers at University College London (UCL) were the first to quantitatively investigate the impact of museum-based collections on wellbeing, specifically the impact on hospital patients of handling archaeological and historical objects (Thomson et al. 2012, Thomson and Chatterjee 2014). The study used quantitative methods to analyse changes in wellbeing over time; results indicated that practical engagement with museums objects had positive impacts on wellbeing. Similarly, the 'Happy Museum Project' suggested that tactile activities that enable 'practical learning' had a positive impact on individual happiness and wellbeing (Happy Museum Project 2011). Research by Garcia and Winn of the NLHF-funded Inspiring Futures Project at the Imperial War Museum North and Manchester Museum provided quantitative longitudinal evidence for positive changes in wellbeing, specifically to individuals from disadvantaged communities involved in the volunteer programme (Garcia and Winn 2017). The research highlighted positive changes in wellbeing that could be directly linked to participatory involvement in the museums: this included improved creativity; aspiration; life satisfaction; social connection; and reduced stress (Garcia and Winn 2017). Furthermore, the research indicated a sustained improvement in wellbeing; 60% of participants noted improved wellbeing over a two- to three-year period and 30% were inspired to secure future education, training, and employment (Garcia and Winn 2017).

A recent quantitative and qualitative evaluation of the Welsh Governments, Welsh Federation of Museums and Art Galleries, and National Museums of Wales Winter of Wellbeing programme examined[27] different festival activities and thirty-eight youth-led community engagement and museum, youth, and community partnership projects that engaged with 4,240 young people and children (Sayer and Luck 2022, Welsh Government 2022). Statistically significant results indicated that this programme improved social, emotional, and physical wellbeing of young people and children, and overall improved wellbeing by 24% (Sayer and Luck 2022).

Quantitative and qualitative wellbeing research has highlighted some of the positive impacts that museum and specific museum programmes can have on wellbeing, which include:

- Personal learning.
- Self-confidence.
- Building friendships.
- Learning new skills.
- Interest in the world around us.
- Learning about selves and world around us.
- Appreciation of own lives.
- Curiosity.

Museum-based wellbeing studies and UCL's Museum Wellbeing Measures Toolkit have provided guidelines and strategies for the museum sector to apply and adapt to assess the impact of museum projects on participant wellbeing (Thomson and Chatterjee 2013, Chatterjee and Camic 2015, Desmarais et al. 2018). A range of individual museum studies, some specifically related to museums and wellbeing, such as 'Spaces for Wellbeing: A Report from The National Alliance for Museums, Health and Wellbeing' have highlighted the benefit of specific projects to improving wellbeing including those with elements of practical engagement, discovery, and volunteering (Desmarais et al. 2018). This research indicated that visiting museums could support the key traits linked to wellbeing, including the opportunity to learn, developing a sense of belonging, ownership, and friendship (Desmarais et al. 2018).

Museums and art galleries impact on wellbeing has been more comprehensively researched than other heritage assets, including heritage sites. As such museums and galleries programmes in the United Kingdom have been able to provide quantitative evidence to align themselves with wider government 'wellbeing' and 'health care' agendas (Cutler 2022). This has included developing projects such as UCL's Museums on Prescription, Holburne Museums Pathways to Wellbeing, Salisbury Museums Well-City Salisbury and Dulwich Picture Galleries Creative Wellbeing, which have been supported by the government and NHS to provide non-medical early preventive intervention (Reily et al. 2018, Desmarais et al. 2018, Hammett and Gascoigne 2022, Campion 2022, Findlay et al. 2022, Ryan 2019, 2022, Cutler 2022). National Museum of Wales Winter of Wellbeing programme sought to link its objectives and outcomes to the Welsh government commitments and wellbeing goals, which included elimination of inequality (including wellbeing inequality) and educational reform (Welsh Government 2022).

The 'Mappiness' project sought to understand the different impacts of specific cultural activities, including museums, theatre, concerts, music, and reading on happiness and wellbeing. Results suggested museums ranked second in happiness and relaxation and that museums had a significant positive effect on wellbeing (Fujiwara and Mackerron 2015). The museums sector has highlighted specific and unique impacts to wellbeing, which include engendering 'trust' specifically in the knowledge presented, the ability of large collections, specialist staff and highly trained volunteers to support learning, the ability of visitors to be supported to learn something new, and providing a safe community space (Phillips 2008, Desmarais et al. 2018, Fujiwara and Mackerron 2015). It is the specific wellbeing values attached to museums, and what makes museums unique and different from other heritage assets, that is of interest to this research.

Museum and Gallery Case Studies

Four case studies from United Kingdom and India formed the basis of research into the impacts of museums and art galleries on wellbeing, these included the Imperial War Museum (UK), Dulwich Picture Gallery (UK), Chandigarh Government Museum and Art Gallery (India), and Venkatappa Art Gallery and Museum (India). The following section provides the results of individual case studies and serves to illustrate contextual and localised nuances in changes to specific visitor wellbeing indicators, resulting from specific experiences at museums and art galleries.

Imperial War Museum Duxford

IWM Duxford is located in the village of Duxford, Cambridgeshire (UK), which has a population of approximately 2,000 and falls within the Liberal Democrat parliamentary constituency of South Cambridgeshire (Office of National Statistics 2011). IWM Duxford is a branch of the Imperial War Museum group, and was established in 1976, as part of a wider national expansion strategy (Asbury 2016). Other branches include IWM London, Churchill War Rooms, HMS Belfast, and IWM North. IWM received just under half its annual funding from DCMS, and is required to support DCMS national priories, and agreed targets and objectives (IWM 2023). The rest of IWM funding is raised from charitable giving, philanthropic support, sponsorship, donations, admission charges, and commercial activities (IWM 2023). Adult ticket prices for day cost £27.50, children 5–15 £13.75, children under 4 free, and concession senior (ages 65) or student £24.75.

IWM Duxford is situated within a former RAF airbase, whose history stretches back to the First World War (Asbury 2016: 7). The site played a significant role in the Battle of Britain during the Second World War. This historic site provides the location for Europe's largest air museum (Asbury 2016: 7). Exhibition halls are located within former aircraft hangers and RAF buildings, which provide the space to exhibit huge planes and tanks (Asbury 2016). Through military aircraft, vehicles, boats, tanks and weapons, aviation and military artefacts, and personal stories of those who served at Duxford and in the RAF, IWM Duxford presents the story of military aviation in Britain, the Commonwealth, and America and of the peoples whose lives have been affected (Asbury 2016).

IWM Duxford attracts about 400,000 visitors a year, and 91% are from the United Kingdom. Of these visitors 64% are male and 36% are female, 3% identify as BAME, 23% consider themselves to have a disability, and 13% of visitors were 65+, 14% were 55–64, 21% were 1–15. Demographics of ninety-seven participants completed both before and after survey indicated that 48% of respondents described their gender as female and 51% as male, 50% were over the age of 50, 90% did not consider themselves to have a disability, 99% described themselves as white, 79% were visiting with family, 90% were not members, and 68% made regular visits to cultural heritage sites. Respondent demographics indicated in comparison of overall visitors' numbers that respondents were skewed towards people describing their ethnicity as white and gender as female, and not having a disability.

Quantitative Analysis

PANAS negative affects visitor scores indicated a reduction in mean scores by an average of –7.39% and nine negative PANAS associated emotions decreased after visiting IWM Duxford (Appendix, Fig. 38). There was a statistically significant decrease in visitor mean PANAS negative scores and reduced variability (standard deviation) in visitors feeling irritable ($p<0.000$), nervous ($p<0.015$), distressed ($p<0.019$), and guilty ($p<0.039$), and reduced but increased in variability for jittery ($p<0.043$) after visiting IWM Duxford (Appendix, Fig. 40). Results indicated improvement in visitors' personal 'emotional' wellbeing, linked to a reduction in negative feelings because of visitor experience at IWM Duxford.

PANAS Positive affects schedule indicated visitor mean scores increased by an average of 11.03% across all measures after visiting IWM Duxford, and all positive PANAS associated emotions improved (Appendix, Fig. 39). Visitors experience statistically significant increases and reduction in variability (SD) in visitor mean PANAS positive scores after visiting IWM Duxford in feeling inspired ($p<0.000$), attentive ($p<0.000$), proud ($p<0.000$), and interested ($p<0.005$), and a statistically significant increase but an increase in variability (SD) in visitors feeling determined ($p<0.000$) (Appendix, Fig. 40). Significant changes in scores indicated IWM Duxford improved personal wellbeing of visitors in the indicators of emotional wellbeing, specifically improvement in positive feelings, positive functioning, specifically improving engagement (attentive, interested), and meaning and purpose (inspired), and self-esteem and resilience (proud).

MWS visitor mean average scores increased by 2.16% and improved for thirteen measures after visiting IWM Duxford (Appendix, Fig. 41). After visiting IWM Duxford, visitors experienced a statistically significant increase and but no change in variability (SD) in feeling happy with life ($p<0.043$) (Appendix, Fig. 42). This indicated an improvement in personal wellbeing, linked to emotional wellbeing and improved happiness.

PANAS and MWS results indicated that visitors to IWM Duxford experienced a measurable and significant improvement to their personal wellbeing as a result of their encounters with heritage, which included: positive functioning as visitors felt more interested, inspired, and attentive as a result of their visit; emotional wellbeing visitors indicated a reduction in negative emotions, including feeling nervous, irritable, distressed, guilty, and jittery and an increase of positive feelings of inspired, attentive, proud, determined, and interested; resilience and self-esteem: visitors indicated they felt prouder due to visiting IWM Duxford. Visiting Duxford supported but did not significantly improve visitors' social wellbeing and had slight negative impacts, although not statistically significant to visitor wellbeing specifically linked to social wellbeing indicators of trust and belonging, as visitors indicated a slight decrease in trusting people. Quantitative results indicate that IWM Duxford supported visitors to keep learning and take notice.

Qualitative Analysis

Thematic analysis of participant commentary indicated that the most meaningful experience included: specific artefacts/exhibitions (predominantly aircraft) (n = 41), spending time and creating connections with family and friends (n = 23), personal stories depicted/connections (including to passed loved ones) (n = 21), enjoyment (n = 13), pride/appreciation/awe/patriotism/heroism (n = 12), education/learning (n = 10), volunteers (n = 3), seeing other visitors enjoying themselves (n = 3), nostalgia (n = 3), and thanks (n = 2). Recurring themes in free comment space included: enjoyment (n = 13), thanks (n = 6), returning (n = 4), suggestions for improvements (n = 4), contemplation (n = 3). Key emotive words used by visitor in free commentary space included: enjoyed, amazing, enthusiasm, inspiring, pride, and excited

Thematic analysis of visitor commentary based on NEF five ways to wellbeing, indicated that visiting Duxford impacted participants in the specific areas, which included to connect, to take notice, to keep learning, and be active. Visitors noted they felt IWM Duxford supported them to connect with each other and supported them to 'spend time with family' (AV). Visitors noted IWM Duxford provided an arena for shared experience and engendered 'a sense of family togetherness—connection' (AV), this facilitated friends and families to connect with each other through sharing memories, with a visitor noting they had 'be[en] reminded of my grandfather's work for the R.A.F. Farnborough, especially during the war with Mrs Shilling and afterword on the Comet disaster investigation' (AV). IWM Duxford enabled visitors to re-connect to their personal memories 'to relive my childhood' (AV) and family histories and supported one visitor to make 'contact with my family who fought and served in the wars and after, I am the son of an RAF engineer' (AV). The exhibitions and experiences at IWM Duxford supported visitors to 'connect with events back in time' (AV) through personal stories told in exhibitions, including 'the back-stories of people relating to the aircraft in the American Air Force' (AV). IWM Duxford supported visitors to take notice and feel 'inspired' (AV) and to them 'think, remember and appreciate' (AV) through 'seeing the amazing things that other people's ingenuity has created' (AV) and the 'heroic efforts by ordinary people, selfless acts' (AV).

Visitors to IWM Duxford commented that through the exhibitions they 'learn[t] about an area of history and the people within it' (AV) and 'gained more knowledge' (AV). Visitors commented IWM Duxford had supported multigenerational learning through 'all generations [being] together' (AV) and their perceived importance of learning together 'being able to learn with my partner' (AV). IWM Duxford supported visitors to keep learning, through sharing and exploring personal family histories, as one visitor commented 'teaching and enjoying historical artifacts which were used by family members no longer with us, keeping memories alive for our 1-year-old son' (AV). Visitors indicated the importance of IWM Duxford in educating future generations and sharing family and community memories down the generation was perceived by one visitor as 'vital education' (AV). Visitors occasionally commented that visiting IWM Duxford and walking around the site supported them to be active, this included commenting that they were 'tired but happy' (AV).

Visitor commentary indicated IWM Duxford impacted personal wellbeing. Visitors noted improvement in the indicators of emotional wellbeing and frequently stated their 'excitement', 'marvel' and 'enjoyment', 'sense of awe and wonder', and that they 'love[d] the whole Duxford experience' (AV). Visitors commented on how 'uplifting [the] experience [was] today' (AV), specifically due to the interactive exhibitions and the intergenerational storied experiences presented. Visitors commented that they 'loved seeing the planes flying, fabulous' (AV). IWM Duxford supported visitors' emotional wellbeing through positive transference of emotions from companions' happiness; 'seeing my hubby's face light up "cos he saw his favourite plane"' (AV) and 'listening to my partner's laughter of enjoyment over my excitedness upon seeing the SR-7' (AV).

Visitor commentary indicated IWM Duxford supported the indicators of reliance and self-esteem. Visitors noted how IWM Duxford, and its exhibitions engendered pride in their family's role in military and world wars, 'I saw a helicopter, a Royal Navy helicopter, that it is likely that my grandfather brought down onto HMS Bulwark as he worked on the deck—I felt very proud' (AV). The exhibitions supported visitors to feel pride and be inspired by the 'selfless acts', 'heroic efforts', and 'peoples' ingenuity' (AV), and that 'seeing the part women played in the war efforts, it was very inspiring, and I feel proud of them' (AV).

Visitors' commentary reflected a positive change in the indicator of positive functioning because of their visit to IWM Duxford, specifically improvement to their enthusiasm and engagement from 'very interesting, very entertaining' (AV) visitors' experiences. Visitors indicated they 'appreciated' (AV) the museum and its ability to create 'an opportunity to connect with events back in time' (AV) and that the exhibitions which included personal stories, such as the American exhibition, were 'thought provoking' (AV) and the Normandy Experience was 'touching and poignant' (AV), and this supported IWM Duxford to be 'a great museum that makes you think, remember and appreciate' (AV). Visitors indicated IWM Duxford had meaning and purpose to them, encouraging visitors to undertake further personal research to understand their family history, for example to find 'out more about how to research our family connection to the Battle of Arnhem' (AV) and to share learning from their visit, 'taking information back to school (primary school teacher) to impart knowledge' (AV). Visitors indicated that IWM Duxford provided them with freedom of choice and autonomy, 'so many stories and experiences, indoors and outdoors—the freedom to be able to journey through it and at our own pace' (AV). Visitors commented that coming to IWM Duxford was an experience that supported their vitality and to be physically active, stating 'I need to come back as I'm exhausted now' (AV).

Commentary indicated that IWM Duxford positively impacted visitors' social wellbeing, including in the components of trust and belonging. Visitors felt a sense of national belonging due to visiting IWM Duxford and the heritage presented there, one visitor noting 'we are very lucky and privileged to live and be a part of this country' (AV). IWM Duxford had positive impacts on wider socio-cultural identity and familial belonging through developing a sense of 'nostalgia related to everything you see when entering the different exhibits' (AV). IWM Duxford helps visitors develop supportive relationships, it provided people to have a place and space to 'spend quality time with the family, and a sense of family togetherness—connection' (AV) and 'spending time with loved ones' (AV), sharing experiences with family and supporting intergenerational interactions through various activities with family, such as 'revisited Concorde where we used to take our now 23-year-old daughter when she was tiny' (AV). Visitors frequently commented that volunteers and staff at IWM Duxford were 'great' (AV), that through these experiences they felt supported to develop new supportive relationships.

Visitors noted some negative impacts on personal wellbeing due to visiting IWM Duxford specifically related to high cost of food, personal experiences, limited access to specific areas, and technical malfunctions with interactives and preservation and conservation of spaces, this included 'the prefab in such a neglected state' (AV).

Qualitative results from visitor commentary indicated that IWM Duxford supported visitors to take notice, connect, be active, and learn, that it was 'a great museum that makes you think, remember and appreciate' (AV). Commentary indicated that visiting IWM Duxford improved personal and social wellbeing. It improved visitors' social wellbeing by providing a social space and place and time to develop supportive relationships and quality interactions with family, friends, and loved ones, and it supported trust and belonging through enabling relationships and memories to be developed with loved ones, and through the respect and sense of belonging engendered between staff, volunteers, and visitors. It improved personal wellbeing, in the category of emotional wellbeing with commentary indicating high frequencies of positive emotions, including feeling happy, relaxed, and calm; positive functioning as visitors felt engaged with history and inspired, that the visit had meaning and purpose, inspiring them to learn, remember and create new memories and they had autonomy. Visitors indicated improved vitality through the active visit to IWM Duxford.

Summary

Triangulation of qualitative and quantitative data indicated that IWM Duxford had a statistically significant and measurable impact on visitors' subjective wellbeing. Visiting IWM Duxford supported visitors in the NAW Indictors of social and personal wellbeing (Michaelson et al. 2009). IWM Duxford improved visitors' social wellbeing specifically in facilitating the development of support relationships with family and friends and improved trust and belonging, specifically people's sense of individual, communal, and national belonging. IWM Duxford improved visitors' personal wellbeing, including positive functioning, measurably supporting engagement and meaning and purpose; emotional wellbeing by supporting visitors to feel more positive, including feeling significantly happier with their life and less negative emotions; resilience and self-esteem as visitors were engendered to feel significantly prouder; and in vitality, as IWM Duxford supporting visitors to feel more active. IWM Duxford measurably supported visitors in the NEF five ways to wellbeing, to be active, to keep learning, to connect, and to take notice of the world around them.

Dulwich Picture Gallery

Dulwich Picture Gallery (DPG) located in Dulwich, South London (UK). It is located in the London Borough of Southwark and in the Labour constituency of Dulwich and Norwood, and has a population of 10,251, of which the majority identify as white (79%) and female (5%) (ONS 2021).

DPG is the world's first purpose built public art gallery, founded in 1811 after Sir Francis Bourgeois RA bequeathed his collection of old masters 'for the inspection of the public' and opened to the public in 1817 (Dulwich Picture Gallery 2023). DPG houses a collection of over six hundred paintings, from European masters including Rembrandt, Rubens, Gainsborough, and Canaletto (Dulwich Picture Gallery 2023). At the time of the survey the gallery was housing the 'Reframed: The Woman in the Window' Exhibition. DPG is a registered charity, and relies of visitor ticket sales, philanthropy, and charitable grants, it is open from Tuesday–Sunday 10 am–5 pm and ticketed with friends and under 18-year-olds free, adults £16.50 and concessions £8 (Dulwich Picture Gallery 2023).

Thirty-four participants completed both surveys, 32% visited on their own, with a large percentage 62% of return visitors. Many participants did not answer demographic questions or undertake qualitative commentary. Demographically participants ages ranged from 20–80, 62% were aged 50+, and 76% described their gender as female; 97% did not consider themselves to have a disability and 87% describe their background as white British. Due to limited demographic data collected a DPG comparison between overall demographic data from visitors to DPG was not able to be undertaken.

Quantitative Analysis

PANAS Negative visitor scores reduced after visiting Dulwich Picture Gallery by an average of –2.25% and decreased for six PANAS negative measures (Appendix, Fig. 43). None of the measurable changes to PANAS negative scores were statistically significant. There was a reduction in variability (SD) reduced for six measures, which included feeling scared, irritable, nervous, jittery, distressed, and guilty (Appendix, Fig. 45). PANAS negative results indicated visitor felt fewer negative emotions linked to improved personal 'emotional' wellbeing after visiting DPG.

PANAS positive visitor scores after visiting DPG indicated an average increase of 15.66% and improvement to all mean scores (Appendix, Fig. 44). Results indicated statistically significant increases and reduced variability (SD) for visitors feeling inspired ($p<0.000$), excited ($p<0.000$), interested ($p<0.006$), enthusiastic ($p<0.027$), and alert ($p<0.049$) (Appendix, Fig. 44). There was a statistical increase and increase in variability (standard reduced) for strong ($p<0.035$) and determined ($p<0.013$) (Appendix, Fig. 45). Significant increases to specific PANAS positive scores indicate that visitor personal wellbeing in the component of positive functioning, and engagement, meaning, and purpose improved due to experiences at DPG and visitors' personal 'emotional wellbeing' improved, and positive emotions and feeling increased due to visiting DPG.

MWS visitor mean scores after visiting DPG increased on average by 6.82% and all mean scores increased (Appendix, Fig. 46). Results indicated that after visiting DPG visitor scores statistically significantly increased and reduced in variability (SD) for opportunity to learn new things ($p<0.000$), sense of accomplishment ($p<0.000$), interest in the world around me ($p<0.004$), satisfied with my life ($p<0.004$) and happy with my life ($p<0.006$), life is worthwhile ($p<0.039$), and overcome challenges ($p<0.012$) (Appendix, Fig. 47). These results indicated that DPG had a measurable and significant improvement to personal wellbeing of visitors, in the components of self-esteem and resilience, satisfying life, positive functioning, and emotional wellbeing.

Quantitative resulted indicate that DPG measurably supported visitors to learn, to take notice, and to connect. It measurably and significantly supported visitors to learn new things and to have the opportunity to learn new things; to take notice, to be more alert, enthusiastic, and be interested in the world around them. DPG measurably and significantly improved visitors' personal wellbeing. Visitors' personal wellbeing was measurably and significantly supported in the subcomponents of emotional wellbeing, which resulted in increased positive feelings and feeling happier with life; satisfying life, as visitor felt more satisfied with life and life was worthwhile; self-reliance and esteem, with people feeling they could overcome challenges; vitality, visitors felt stronger; and positive functioning, with visitors feeling interest in the world around and sense of accomplishment. There was no statistically significant impact on social wellbeing because of visiting the gallery.

Qualitative Analysis

Thematic analysis of visitor commentary indicated common and recurring themes in visitor perceptions of the most meaningful experience whilst at DPG including: calming relaxing experience (n = 3); inspiration (n = 2), engagement/interest (n = 10), education (n = 2), time with family/friends (n = 4), improved wellbeing (n = 11) and art/exhibitions (n = 10). Analysis of free comment space indicated key recurring emotive words were time (6), exhibit (5), feel (5), interested (5), ages (4), mad (4), seeing (4), women (4), art (3), look (3), spend (3), window (3), and yes (3). Key emotive words used by visitor in free commentary space included calming, relaxing, interesting, family, and friends.

Visitor commentary indicated that DPG impacted NEF five ways to wellbeing, to learn, to connect, and to take notice. Participants frequently used the word 'interesting' or 'interest' and comments from visitors including how DPG had 'educated me' (AV), 'opened up my eyes and mind to new perspectives and ways of looking', and 'he loves the clocks and learning about how they work' (AV) indicated visitors had learnt new things. Commentary reflected the impact DPG had on supporting visitor to connect with both people and art; visitors noted they had connected with their family and friends 'spending time with family made things seems a little brighter' (AV) and spent 'a wonderful time with my son' (AV). Visitor comments reflected the development of new and increased connection to the artwork: 'reading up on the captions made me feel more connected to the surrounding artwork' (AV) and a visitor noted DPG had supported them with 'reconnecting with my interest in painting' (AV). Comments indicated that through their engagement with 'the great exhibition[s]' (AV), the 'variety', and seeing the 'skill of the works' (AV) visitors felt supported to take notice of the world around them, specifically through art. Visitors commented that their experience at DPG resulted in them feeling 'inspired' (AV), how 'it was interesting to look at [and] appreciate the level of detail' (AV), and that it was 'intellectually stimulating, it made me think' (AV).

Visitor comments indicated DPG supported personal wellbeing and social wellbeing. In relation to personal wellbeing, DPG improved visitors' emotional wellbeing; common words used by visitors included 'enjoy', 'calm', 'inspired', 'relaxing', and 'uplifting'. Visitor comments such as 'I've enjoyed it' (AV), 'it made me happy' (AV), and 'calming [and] relaxing experience' (AV) and 'I always feel uplifted and inspired after visiting an art gallery' (AV) indicated the positive impact of visiting the DPG had to emotional wellbeing.

Comments indicated that visiting DPG impacted personal wellbeing category positive functioning, supporting visitor engagement through 'stimulating -great art' (AV) that 'inspired' (AV) the visitor. Visitors indicated high levels of 'interest' and that they felt more 'alert' (AV). Photographs and artwork supported visitors to engage with both the art, other viewpoints, and wider social cultural debates, providing meaning and purpose to the visit, and to consider 'more about the sense of woman through the ages' (AV) and provided 'time to think about other people's experiences shown in the artwork' (AV). A visitor indicated that DPG had enabled them a sense of autonomy to 'to do my own without my kids' (AV).

Commentary indicated DPG supported visitors' social wellbeing, specifically helping develop and solidify supportive relationships through 'spending time with friends' (AV) and 'spending time with family' (AV). Visitors commented on how coming with their children had positively impacted their relationships and wellbeing, 'spending time with my son is always a joyful experience' (AV), and 'I have spent a wonderful time with my son' (AV).

Qualitative analysis provided evidence that DPG supported visitors to learn, to take notice, and to connect. Visiting the DPG improved visitors' subjective wellbeing, specifically in the components of personal wellbeing, in the area of emotional wellbeing with high frequencies of positive emotions and positive emotional comments by the visitors, including relaxing, calm, and uplifted; positive functioning with visitors feeling more engaged with art and culture after their visit, the exhibitions had meaning a purpose to them and wider society and supported them to learn and provided them with autonomy. DPG supported visitors' social wellbeing, specifically helping maintain and build supportive relationships with family and friends.

Summary

The triangulation of qualitative and quantitative results indicated that DPG had measurable and significant positive impacts on visitors' subjective wellbeing. It provided visitor experiences that support the NEF five ways to wellbeing supported visitors to connect, keep learning, and to take notice of the world around them. DPG positively impacted the NAW indicators, supporting visitors' personal wellbeing and social wellbeing. DPG measurably and significantly improved personal wellbeing, this included: emotional wellbeing, as visitors' positive feelings of happiness increase; positive functioning, as visitors felt more interested, inspired, alert, excited, and enthusiastic; satisfying life, with visitors feeling more satisfaction with their lives; self-esteem and reliance, as visitors felt more determined and an increased sense of accomplishment, and felt more able to overcome challenges. DPG measurably improved social wellbeing; specifically it provided a space, place, and time to develop supportive relationships with family and friends and feel connected to other people and the art around them.

Chandigarh Government Museum and Art Gallery

Chandigarh Government Museum and Art Gallery (CGMAG) is in located in the city of Chandigarh, Uttar Pradesh, India. Chandigarh is currently governed by the Bhartiya Janata Party and has an estimated population of approximately 1,325,000 (Census 2011). CGMAG is currently managed by the Uttar Pradesh administration. As such it is funded by local government through public funds, ticket sales, and philanthropic donations; it does not seek to make a profit. CGMAG was established in 1947 to provide a space to present Indian history and the history of Partition (Chandigarh Government Museum and Art Gallery 2020). The modernist rectangular building suspended on stilts was designed by Le Corbusier as part of wider town planning of Chandigarh and opened in 1968 to provide not only a museum 'exhibition space', but also a throughfare, to enable movement of the public in and around the space, and open plan 'galleries' (Chandigarh Government Museum and Art Gallery). CGMAG currently houses the collection of 13,000 Gandharan sculptures, Pahari and Rajasthani miniature paintings, and decorative arts and objects, including modern art (Chandigarh Government

Museum and Art Gallery 2020). CGMAG has galleries specifically related to sculpture, miniatures, decorative objects, and a children's art gallery. Much of the material now housed in the museum was in the museum of Lahore before Partition (Chandigarh Government Museum and Art Gallery 2020). The museum displays focus on specific objects, artefacts, and works of art with limited interpretative narrative. Visitors can either take a self-guided or guided tour (although this is limited to specific periods and days). The museum is open to the public from 10:00 am–5:00 pm Tuesday to Sunday, and admission charges for individuals over the age of 18 are INR 30 (approximately 30 pence), but it does not charge children under 18 or school groups. There is a higher price for foreign and non-resident visitors.

Seventy visitors completed both before and after surveys: 67% were aged 20–29, 63% were female, 95% did not consider themselves to have a disability, and 60% were first-time visitors. There is demographic data available from CGMAG to compare survey demographics.

Quantitative Analysis

PANAS negative visitor scores after visiting CGMAG decreased by an average of −5.25% across all scores and all mean scores decreased (Appendix, Fig. 48). Visitor PANAS negative scores after visiting CGMAG statistically significantly decreased and variability (SD) reduced for upset ($p<0.010$) and scared ($p<0.027$), and visitors felt statistically significantly less jittery ($p<0.040$), although variation (SD) increased (Appendix, Fig. 50). Measurable reduction in PANAS negative scores and significant reductions to some negative feelings indicated visitor personal 'emotional' wellbeing improved after visiting CGMAG.

PANAS positive visitor mean scores after visiting CGMAG increased by an average of 7.79% and mean scores increased across nine measures (Appendix, Fig. 49). Statistically changes to visitors' mean scores after visiting CGMAG indicated significant increase, yet also increase in variability (SD) in interest ($p<0.000$), inspired ($p<0.003$), proud ($p<0.007$), excited ($p<0.012$), enthusiastic ($p<0.037$), and active ($p<0.037$) (Appendix, Fig. 50). Results suggested visiting CGMAG measurably improved visitors' personal wellbeing, specifically to emotional wellbeing in improved positive feelings, positive functioning in the components of meaning and purpose and engagement (interested, excited, inspired, enthusiastic), self-esteem and reliance (proud), and vitality (active); increased variability indicates caution interpretating the statistical significance of the results.

MWS visitor mean scores after visiting CGMAG increased across all measures by an average of 6.32% (Appendix, Fig. 51). Statistically after visiting CGMAG visitors' mean scores indicated significant increases and reduced variability (SD) in opportunity to learn new things (p<0.000), satisfied with life (p<0.009), trust people (p<0.009). Visitors' mean scores indicated significant positive change in life is worthwhile (p<0.009) but increased SD (variability) (Appendix, Fig. 52). MWS results indicated that visiting CGMAG measurably and significantly improve personal wellbeing in the indicators of positive functioning (meaning and purpose), life satisfaction, and improved social wellbeing in the indicator of belonging and trust.

Visiting CGMAG had slight measurable negative impacts, although these not statistically significant to visitor wellbeing specifically feeling guilty and strong, indicating some negative impact on positive functioning and personal wellbeing.

Quantitative analysis indicated that visitor experiences at CGMAG resulted in measurable and significant improvement and change to visitors' personal and social wellbeing. In the indicator of personal wellbeing, visitors' wellbeing measurably and significantly improved in the categories of positive functioning, as visitors' changes to scores indicated the visit had meaning and purpose as they learnt new things; emotional wellbeing, with a reduction in visitors' negative emotions, with visitors feeling less upset and scared after visiting and an increase of positive feelings; satisfying life visitors felt more satisfied with life. Results indicated visitors' social wellbeing was measurable and significantly improved, specifically in the component of trust and belonging as visitors felt supported to trust people. Visiting CGMAG measurably supported visitors to learn, to connect, and to take notice.

Qualitative Analysis

Thematic analysis of the visitors' commentary was limited due to only eighteen participants completing free space commentary on surveys and few comments within most meaningful experience. Key emotive words used by visitors in free commentary space included learning, experience, and beauty. There were no negative comments from the participants.

Thematic analysis based on NEF five ways to wellbeing, indicated that visiting CGMAG impacted participants in the arenas of to keep learning and to take notice. Visitors indicated that visiting CGMAG supported them to keep learning, commenting that it 'was a good learning experience' (AV) and 'the whole visit was knowledgeable' (AV). Visitor comments indicated that learning was specifically focused on historical narratives relating to miniature 'painting' and 'sculptures' (AV). Visitors commented that CGAGM was a 'great experience' (AV), that supported them through 'unheard stories' (AV) and 'insight[ful] stories' (AV) to take notice of world around them.

Visitor commentary indicated that experiences of object and stories relating to objects at CGAGM impacted visitor personal wellbeing, specifically emotional wellbeing, with visitors commenting 'the Shah Jahan Painting made my day' (AV), 'it was a good experience' (AV), and 'artifacts were beautiful' (AV). Commentary suggested CGMAG both as a place and the items presented with supported positive functioning, specifically supporting engagement, 'the miniature paintings really intrigued me' (AV) and 'the architecture of the museum is quite fascinating', and supporting meaning and purpose, with a visitor describing CGMAG as 'quite a meaningful experience, learnt new things' (AV).

Qualitative analysis of visitor commentary suggested that visiting CGMAG positively impacted visitors' personal wellbeing, specifically emotional wellbeing and positive functioning (meaning and purpose and engagement) and supported visitors to take notice and to keep learning.

Summary

The triangulation of quantitative and qualitative results indicated that CGMAG positively impacted the NAW indicators supporting visitors' personal wellbeing and social wellbeing. CGMAG measurably and significantly improved visitor personal wellbeing, specifically emotional wellbeing, and positive functioning, and satisfying life and social wellbeing, specifically trust and belonging. CGMAG provided visitor experiences that support the NEF five ways to wellbeing supported visitors to connect, keep learning, and to take notice of the world around them.

Venkatappa Art Gallery

The Venkatappa Art Gallery (VAG) is situated in Bangalore, India, in the vicinity of Cubbon Park and next to the Bangalore Government Museum and the Visvesvaraya Industrial and Technological Museum (Department of Archaeology, Museums, and Heritage 2023). Bangalore is a city located in the southern Indian state of Karnataka, with a population of over 10 million, and currently governed by the Bhartiya Janata Party (Government of Karnataka 2023). VAG is owned, managed, and funded by the regional Government of Karnataka Department of Archaeology and Heritage; it does not seek to make a profit. VAG was founded in 1966, when the government of Mysore decided that a gallery was required to house the paintings, musical instruments, and plaster of Paris bas reliefs that made up the collection of one of Karnataka's most famous artists, K. Venkatappa (Department of Archaeology, Museums, and Heritage 2023). In 1967, the Chief Minister S. Nijalingappa laid the foundation stone for the Venkatappa Art Gallery and in 1975 VAG was completed. Initially it displayed K. Venkatappa's art and provided space for Karnataka's artists to present contemporary art (Department of Archaeology, Museums, and Heritage 2023). In 1993, the painter K. K. Hebbar donated his collection to VAG and these and others purchased by the Karnataka government including the collections of sculptor Rajaram are displayed at VAG (Department of Archaeology, Museums, and Heritage 2023). VAG is still incomplete; only three of the five floors are in use. VAG is open to the public Tuesday–Sunday from 10:00 am to 5:30 pm, and the cost of a ticket is INR 10 (approximately 10 pence) for adults, with higher prices for foreign visitors and non-residents.

Sixty-four visitors completed before and after surveys at VAG. Demographics indicated 70% were aged 20–29, 61% were female, 100% had no disability, 57% were visiting with friends, and 57% were regular cultural visitors. There were no visitor demographics from Venkatappa Art Gallery available for comparative analysis.

Quantitative Analysis

PANAS negative visitor results indicated after visiting VAG mean scores reduced across all measures by an average of –8.49% (Appendix, Fig. 53). After visiting VAG, visitors' mean PANAS negative scores statistically significantly decreased and reduced in variability (SD) for feeling guilty ($p<0.043$), and there was significant decrease and increased in variability (SD) in visitors feeling hostile ($p<0.031$) and upset ($p<0.031$) (Appendix, Fig. 55). PANAS negative results indicated that VAG had measurable and significant positive impact on visitors' personal 'emotional wellbeing', by reducing negative feelings and emotions.

PANAS positive affects visitor mean scores after visiting VAG indicated increased for all measures by an average of 17.02% (Appendix, Fig. 54). After visiting VAG, visitor mean scores statistically significantly increased for feeling attentive ($p<0.000$), determined ($p<0.000$), inspired ($p<0.000$), interested ($p<0.000$), proud ($p<0.000$), active ($p<0.001$), excited ($p<0.001$), enthusiastic ($p<0.001$), and strong ($p<0.029$), yet only interest reduced in variability (Appendix, Fig. 55). Results indicated VAG improved visitor personal wellbeing in the component of positive functioning (engagement) and emotional wellbeing as visitor positive emotions and feelings increased.

MWS visitor mean scores after visiting VAG increased by an average of 5.90% and increased across thirteen scores (Appendix, Fig. 56). After visiting VAG, mean visitors' scores statistically significantly increased and variability (SD) reduced for feeling life was worthwhile ($p<0.031$), and statistically significantly increased but variably did not change for connected to people ($p<0.002$) (Appendix, Fig. 57). After visiting VAG, visitors' mean scores statistically significantly increased but variability increased for feeling optimistic about my future ($p<0.000$), satisfied with my life ($p<0.001$), sense of accomplishment ($p<0.001$), physically active ($p<0.026$), and freedom of choice ($p<0.045$) (Appendix, Fig. 57). MWS results indicated that visiting VAG measurably and significantly improved visitors' personal wellbeing in the component of self-esteem and reliance (life is worthwhile), and in social wellbeing, in the component of supportive relationships (connected with people).

The visitor experiences at VAG significantly and measurably improved their personal wellbeing in the components of emotional wellbeing, supporting visitors to feel more positive and reduced negative feeling; positive functioning, with visitors noting improvement in indicators engagement in their interest; self-esteem and reliance was improved with visitors feeling life was more worthwhile. Visiting the museum had significant impacts on the components of social wellbeing, specifically supportive relationships as visitor felt more connected to people. Quantitative results indicated that visiting measurably and significantly supported users to be connected and to take notice of the world around them.

Qualitative Analysis

Thematic analysis of the visitors' comments indicated that most meaningful experience during visit was the art, with key words included art/arts, good, nice, and place additional key words emotive words included experience, inspired, and beautiful. Visitor comments frequently related their experience to specific objects including art, paintings, and sculptures.

Thematic analysis visitor comments based around the NEF five ways to wellbeing indicated visiting VAG supported visitors to connect, keep learning, and take notice. Visitors noted the impact that VAG physical environment and place had on learning: this included 'good environment and new things to see and learn about' (AV) and 'I'm an art student from Delhi I had learnt about art but after coming here I saw many arts which I didn't learn' (AV) and 'nice place and good environment, learn a lot in the place and good to see such things kept and preserved' (AV). Visitors commented that VAG helped them 'explore more arts' (AV) and provided a 'good place, got to see many things and learn few things inside the gallery' (AV). Visitors noted the presentation of artefacts and painting physical space provided them with a 'good' (AV) experience and 'very nice place to see beautiful paintings and ancient sculptures carved in stone' (AV) and view 'historical artifacts and different drawings' (AV) that supported them to take notice of 'new things around me' (AV). Some visitors commented that VAG provide a space and place that 'suitable to spend time with family and friends' (AV) and supported them to connect with friends and family.

Visitors' comments indicated that VAG supported the NAW indicators. Commentary indicated VAG improved personal wellbeing and social wellbeing. In relation to personal wellbeing, commentary indicated VAG supported visitor emotional wellbeing by increasing positive emotions, for example visitors noted they had 'been inspired a lot by artwork' (AV) and 'inspired by his works' (AV). Visitor made specific reference to wellbeing benefits including how VAG was a 'calm place to bind your emotions' (AV) and had engendered feelings of peace and love: 'I loved this place, the art collection is good but not huge, but this place is so much peaceful. Anyone who is an art lover will definitely fall in love' (AV). Visitors' comments suggested VAG had a positive impact on their positive functioning, as engagement was enabled through 'exploration' (AV) and it was 'good place for all artists to get inspired' (AV), that supported 'more interest on art or sculpture' (AV). Commentary indicated the 'meaningful arts on the wall are perfect' (AV) at VAG, and provided a meaningful experience for visitors that 'satisfies' (AV). VAG improved visitors' social wellbeing; it provided space and place to 'spend time with family and friends' (AV) and develop supportive relationships.

Visitors' commentary indicated the negative impacts VAG on personal wellbeing, specifically on positive functioning. Negative impacts of VAG on positive functioning were associated by visitors to depth of learning and detail in material presented to visitors; comments included that interpretation was 'very superficial' (AV), there was 'not much description is written' (AV), and 'some of the description is not in English' (AV). A visitor suggested lack of detailed interpretations was why 'people are not enthusiastic to go there' (AV) and did not engage with VAG. Other negative impacts on wellbeing included VAG physical space and buildings being 'poor[ly] maintained' (AV), and it was noted 'the vicinity can be more neatly maintained' (AV) and it had 'no proper lighting' (AV).

Thematic analysis of visitor commentary indicated that visiting the VAG improved visitors' personal wellbeing in the components of emotional wellbeing, as material presented at VAG supported visitors to have positive feeling and emotions, and positive functioning, as viewing art and historic objects supported visitors to engage in the world around them and had meaning and purpose by supporting visitors to learn. VAG improved visitors' social wellbeing in providing a place and space to develop supportive relationships.

Summary

The triangulation of quantitative and qualitative results indicated that VAG positively impacted the NAW indicators supporting visitors' personal wellbeing and social wellbeing. VAG measurably and significantly improved visitor personal wellbeing, specifically emotional wellbeing, positive functioning (meaning and purpose, and engagement), self-esteem and resilience, and social wellbeing, specifically supportive relationships. VAG provided visitor experiences that support the NEF five ways to wellbeing, supporting visitors to connect, keep learning, and to take notice of the world around them.

Comparison between Museums and Art Galleries

Comparison of visitor quantitative and qualitative data from museum and art gallery case studies revealed variation between places and specific impacts to visitors' subjective wellbeing indicators based on geographic locality, nature of visitor experiences, and type and age of material presented, for example archaeological material, historical material, sculpture, and painting.

PANAS negative mean results indicated that visitors to all case study sites experienced a measurable reduction in negative feelings and emotional wellbeing measurably improved at all museums and galleries. Visitors to IWM (–7.4%) and VAG (–8.49%) experienced the largest reduction in PANAS negative mean scores after visiting, and visitors to both places indicated statistically significant reductions feeling guilty (Appendix, Fig. 58). Visitors to IWM experienced the largest (four) number of statistically significant reductions to PANAS negative mean scores after visiting, including uniquely to distressed, irritable, and nervous. After visiting CGAGM, visitors experienced statistically significant reductions in feeling upset and scared, whilst after visiting DPG visitor did not experience any significant reductions to PANAS negative scores. These results suggested visitor experiences at IWM, the only museum in the study located within a heritage site (a former RAF airfield), had the greatest statistically significant impact to reducing negative emotions, linked to improved emotional wellbeing.

PANAS positive mean scores indicated after visiting case study museums and galleries visitors' emotional wellbeing improved; their positive feelings and emotions improved after visiting. VAG (15.63%) and DPG (12.71%) had the greatest positive impact on PANAS positive visitor mean scores, followed by IWM (11.02%) and lastly CGMAG (7.79%) (Appendix, Fig. 58). The average mean percentage increase in PANAS positive visitor scores and statistical significance of specific PANAS positive measures varied between case studies.

PANAS results indicated the museums and art galleries improved visitors' subjective hedonic wellbeing, by positively affecting their positive feelings and reducing negative feelings associated with happiness (Fig. 5.1). They indicated VAG had the most optimal impact on hedonic wellbeing 'happiness' and had greatest impact on reducing negative feelings and improving positive feelings; this was closely followed by IWM (Fig. 5.1).

Visitors to DPG experienced the highest number of statistically significant increases to PANAS positive mean scores, including to inspired, excited, interested, enthusiastic, and alert (5) (Appendix, Fig. 45). After visiting, visitors to IWM indicated a statistically significantly improvement to PANAS positive scores to feeling inspired, attentive, proud, and

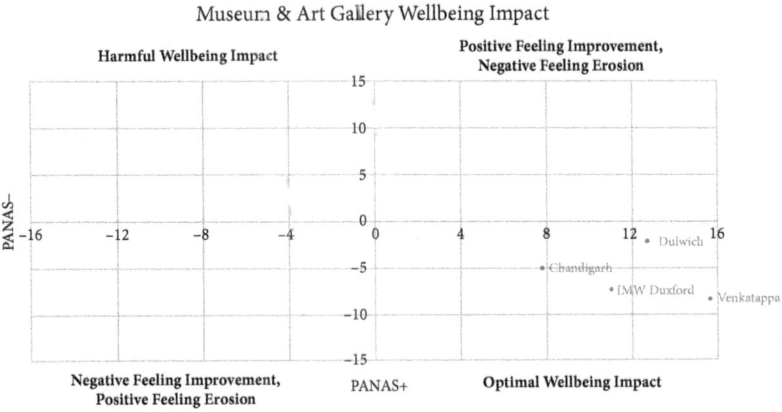

Fig. 5.1 Mapping of museums and galleries PANAS scores relating to optimal hedonic wellbeing impact.

interested (4); visitors to VAG indicated a statistically significant increase to interested (1); and visitors to CGMAG experienced not statistically significant changes (Appendix, Figs 40, 50, and 55). IWM, DPG, and VAG all statistically significantly improved visitors' personal wellbeing. IWM, DPG, and VAG had significant impacts to visitors' positive functioning, with all visitors feeling improved interest. IWM and DPG improved visitors feeling inspired; DPG excited, entrusted, and alert; and IWM attentive. Only IWM significantly improved visitors' self-esteem, as visitors felt significantly prouder after visiting. PANAS positive results indicated that VAG, DPG, and IWM supported visitors to take notice. UK museums and galleries (IWM/DPG) had the most significant impact on improving visitors' positive feelings related to happiness and their personal 'emotional wellbeing', whilst museums and galleries in India had lower levels of significant impacts to visitors' personal wellbeing. PANAS positive scores indicated that visitors to art galleries (DPG and VAG) had the most measurable impact on personal emotional wellbeing, potentially indicating the unique impact of art on visitors' subjective wellbeing and feelings of happiness.

Visiting DPG (7.09%) and CGMAG (6.29%) had the most impact on visitors' MWS mean scores, followed by VAG (5.85%) and IWM (2.16%) (Appendix, Fig. 58). The categorization of museums and galleries MWS visitor mean average scores based on questions linked to social and personal wellbeing indicators enabled the comparison between case study sites based on their specific impact on visitors' personal and social wellbeing indicators. Results indicated CGMAG had the highest impact on visitors' social wellbeing and DPG on visitors' personal wellbeing, with both places having most optional impact on visitors' eudaimonic wellbeing (Fig. 5.2), whilst IWM had the lowest levels of impact to visitors' social and personal wellbeing indicators (Fig. 5.2). Results indicated museums and galleries had the most measurable impact on personal wellbeing indicators (Fig. 5.2).

Visitors to DPG experienced the highest number of mean scores that statistically significantly increased and reduced in variability after visiting, which included uniquely supporting visitors to overcome challenges. Visitors to IWM did not experience significant changes or reductions in variety to any MWS scores, although there was a significant improvement in happy with life, linked to emotional wellbeing. DPG and CGMAG had significant impact on supporting visitors to learning new things linked to positive functioning and to life satisfaction linked to wellbeing indicator satisfying life. DPG and VAG significantly impacted visitors feeling life

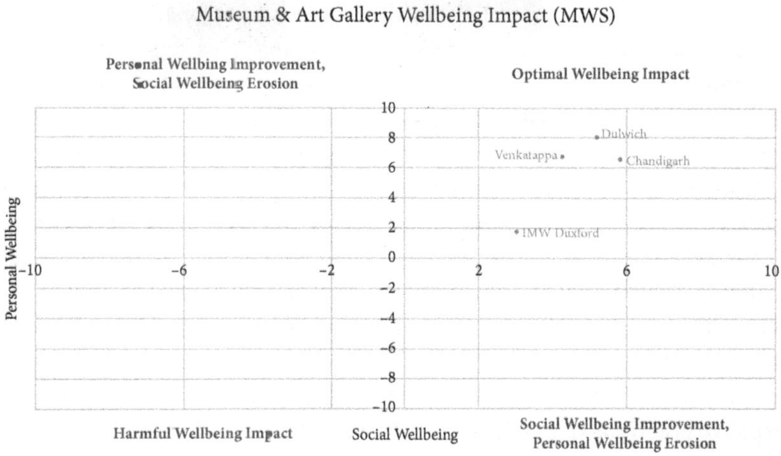

Fig. 5.2 Museums and galleries wellbeing impact based on MWS social and personal wellbeing indicator scores.

was worthwhile linked to self-esteem and resilience, yet only DPG improved visitors feeling able to overcome challenges. Only CGMAG significantly increased visitors' social wellbeing, specifically in the indicator of trust and belong linked to improved trust in people.

Comparison of case study MWS results indicated all museums and art galleries impacted personal wellbeing, yet also suggested they affect different indicators including positive functioning and satisfying life, and only CGMAG, the Indian gallery and museum, had measurable and significant impact on social wellbeing. Visitors to DPG and CGMAG experienced significant changes to learn new things linked to NEF component to keep learning, and participants visiting CGMAG and DPG supported visitors to take notice, but only CGMAG supported visitors to connect.

Museum and gallery case studies all significantly impacted NAW indicators for wellbeing. This included personal wellbeing, in the components of positive functioning and emotional wellbeing. DPG had most significant impact on personal wellbeing, across a broad range of indicators including emotional wellbeing, positive functioning, satisfying life and self-esteem, and IWM had the most significant impact on emotional wellbeing.

CGMAG was the only museum and art gallery that positively impacted social wellbeing, specifically trust and belonging, and supported visitors to connect. This museum specifically related to Indian history focusing on national narratives and independence indicated the important of narratives of belonging and those that relate to personal, familial, and national narratives in supporting social wellbeing. This suggest that visiting museums or art galleries that present local and national narratives that the visitor associates with and is linked to wider national memory and identity had the greatest impact improving social wellbeing, by enabling visitors to feel a sense of belonging and trust. Quantitative results suggested that CGMAG, a place that was both a museum and art gallery that presented archaeological, historical, and contemporary objects and artistic material, provided visitor experiences that most successfully supported NEF's five ways to wellbeing, and improved participants SWB and EWB.

Results demonstrated that IWM, a place that presented historical items located on a heritage site, had the most significant impact on reducing negative emotions of participants after visiting. Art galleries of DPG and VAG that presented both art history and modern art had the most significant impact on improving the positive emotions of participants after visiting. Visiting DPG, an art gallery in the United Kingdom, resulted in the highest number of visitors' personal wellbeing indicators being improved, whilst art galleries and museums in India had the least impact on visitors' personal wellbeing.

Thematic analysis of qualitative data from the case study museums and galleries reflected key differences on their impacts on visitors' personal and social wellbeing. These differences were based on the geographical context place and on the nature of material presented within the space. Analysis of visitor commentary indicated that IWM had the greatest impact on visitors' social and personal wellbeing indicators. Visitor commentary from CGMAG indicated the low levels of change to personal wellbeing indicators and did not reflect any social wellbeing impacts, unlike the quantitative results.

Qualitative analysis suggested that case study sites impacted different personal wellbeing indicators. Visitors to DPG, VAG, and IWM indicated that they had a new 'experience' and a key word in visitor commentary at these sites was 'inspired' (AV). At these sites visitor commentary frequently stated an interest and engagement with a specific historical object or historical narrative. This indicated that visiting DPG, VAG (galleries), and IWM (museum) impacted visitors' personal wellbeing, in the component of positive functioning and by visiting these heritage spaces visitors felt supported to take notice of the world around them. In contrast to the quantitative data, qualitative data indicated that visitors to all case study museums and galleries frequently felt supported to learn and 'learn new things' (AV), with learning and education appearing as key words in visitor commentary.

Visitors to case study sites in India (VAG and CGMAG) indicated a particular impact on improving visitors' emotional wellbeing, which was based on the sites perceived 'beauty' by the visitors, and subsequently improved positive feelings. At IWM visitor commentary indicated that this site had a unique impact on visitors' personal wellbeing in the component of vitality and indicated that the visit supported them to be active.

Qualitative analysis indicated variation between museums and galleries on their impact on visitors' social wellbeing. Visitor commentary from IWM, DPG, and VAG indicated that visits had supported visitors' social wellbeing, specifically in developing and creating new supportive relationships. Visitors frequently noted that these museums had provided places and spaces to develop relationships and to spend time with family and friends. After visiting IWM, visitors were unique in making specific comments regarding improvement in belonging and trust related to the development of personal memories and personal connections linked to familial and personal identity. In contrast to the quantitative data, in which CGMAG was the only site that had measurable impact of social wellbeing, visitor commentary on CGMAG did not reference social wellbeing impacts.

Qualitative analysis indicated that all museums and galleries in this study improved visitors' personal wellbeing and visitors felt experiences had supported their personal wellbeing in the components of emotional wellbeing and positive functioning. Qualitative results indicated that all museums and galleries in this study supported visitors to take notice and to learn. Visitors to IWM, DPG, and VAG noted improvement to their social wellbeing after visiting these sites. Qualitative data indicated that visitors to IWM, a museum which is situated at a former RAF airbase and involved indoor and outdoor exhibitions and interactive displays experienced the broadest qualitative impact to their personal wellbeing (emotional wellbeing, positive functioning, vitality) and social wellbeing (supportive relationships and belonging and trust) indictors. Visitor commentary at this site also reflected the strongest sense of belonging and connections to heritage based on individual memory and identity.

Commentary reflected national differences in expressions, commenting, and providing written feedback. For example, commentary at Indian sites, as due to Covid international travel restriction all visitors were Indian or Indian residents, was shorter and most visitors did not write in free commentary space. At Indian museums and galleries visitors focused and appeared more comfortable with the quantitative elements of survey. When visitors to Indian sites did make comments, these were less empathic or personal and made limited reference to social wellbeing impacts, specifically trust and belonging linked to memory or individual identity.

Conclusion

Triangulation of quantitative and qualitative results indicated that visitor experiences at museums and galleries impacted their subjective personal wellbeing and social wellbeing. It was the differences in components in which these wellbeing indicators and hedonic and eudaimonic wellbeing capabilities were influenced that enabled patterns to be seen in types of material presented and in the context of the museum and gallery. This included the influence of the nature of the material presented on personal, community, and national identity and memories and the impact this had on social wellbeing. Results suggested museums, such as IWM and CGMAG, which presented stories that triggered personal memories and linked to national and local identities had the greatest impact on visitors' subjective personal and social wellbeing, and uniquely within this group of case studies supported visitors to connect and improved visitor eudaimonic wellbeing. Galleries that presented creative 'artistic' material enabled people to emotionally connect and improved personal wellbeing, specifically emotional wellbeing linked to improvement in happiness and hedonic wellbeing, and positive functioning linked to improvement in eudaimonic wellbeing. Art galleries supported visitors to keep learning and to take notice, whilst museums supported visitors to connect.

The results from this study have demonstrated how differences between museum and gallery interpretative practices and approaches to visitor experiences impact their ability to support visitors' social wellbeing, specifically to support visitors to connect and develop trust and belonging. Differences related to the role of the volunteers and staff in interpretative practices and the nature of the displays. Supporting social wellbeing was most successful at museums where visitors personally connected to the material presented or to the people presenting the material. For example, at IWM the intergenerational empathic and emotive first-person stories and oral history testimonies, both in interpretative material and through volunteer and staff storytelling, resulted in visitors experiencing the greatest emotional connection and highest levels of emotional wellbeing.

Research demonstrated that whilst all museums and galleries supported visitors to take notice and to keep learning, it was museums that presented broader national histories, such as IWM and CGMAG, that successful supported visitors more broadly in four of the five NEF ways to wellbeing including to connect, to be active, to take notice, and to keep learning.

Heritage and Wellbeing: The Impact of Heritage Places on Visitors' Wellbeing. Faye Sayer, Oxford University Press.
 DOI: 10.1093/9780191914539.003.0005

6
Heritage Type, Demographics, and Context

Visitor Wellbeing at Heritage Places

This chapter analyses the difference between diverse heritage places including heritage sites, archaeological sites, museums, and art galleries and examines the range of impacts each place has on subjective wellbeing. Comparative analysis will identify trends in capabilities of subjective visitor wellbeing impacted by different types of heritage place, seeking to uncover how specific types of heritage places impact hedonic and eudaimonic wellbeing, and social and personal wellbeing indicators. The chapter examines how the national context in which heritage places are situated impacts on their potential to improve visitors' subjective wellbeing and uncovers how organizational practices and interpretative approaches applied at heritage places can impact visitors' wellbeing. The chapter provides evidence of how visitors' age, gender, or disability impacts heritage places' ability to improve their subjective wellbeing. It considers if demographic patterns in the wellbeing impact of heritage places can be identified across national boundaries. It provides empirical evidence to help organizations to understand the role of heritage places nationally in improving individuals' subjective wellbeing and considers the potential of these sites to tackle societal wellbeing inequalities, related to age, gender, and disability.

Museums' and Art Galleries' Impact on Wellbeing

Visitors from four case studies of museums and art galleries in the United Kingdom and India completed before and after surveys. Demographically, respondents identified their gender as 56% female, 38% male, and 6% preferred not to answer the question; 5% of respondents indicated they had a disability, 87% did not have a disability, and 8% did not answer the question; 3% were under the age of 20, 43% aged 20–29, 15% aged 30–39, 9% aged 40–49, 11% aged 50–59, 15% were over 60, and 4% did not answer. Demographically, the typical visitor was aged 20–29, female, and did not have a disability.

After visiting museums and galleries, the PANAS negative mean average scores of these 263 visitors decreased by 6.5% (Appendix, Fig. 59) and statistically significantly decreased and reduced in variability (SD) for feeling irritable ($p<0.000$), scared ($p<0.001$), jittery ($p<0.003$), upset ($p<0.006$), afraid ($p<0.030$), and ashamed ($p<0.043$) (Appendix, Fig. 60). After their visits to museums and art galleries, their PANAS negative mean scores statistically significantly reduced and variability increased for feeling nervous ($p<0.006$), distressed ($p<0.020$), and guilty ($p<0.036$) (Appendix, Fig. 60). Visitor PANAS positive mean average scores increased by 11.8% and scores statistically significantly increased and reduced in variability (SD) for feeling interested ($p<0.000$) and attentive ($p<0.000$) (Appendix, Figs 60 and 61). PANAS results indicated visiting museum and galleries resulted in measurable and significant improvement in visitors' personal wellbeing, in the components of emotional wellbeing, linked to a reduction in negative feelings, and an increase in positive feelings and positive functioning, together with improvement in engagement, meaning, and purpose.

After visiting museums and galleries, MWS results indicated visitor average mean score across all measures increased by 4.7% (Appendix, Fig. 62). Visitors MWS scores statistically significantly increased and reduced in variability (SD) for opportunity to learn new things ($p<0.000$), sense of accomplishment ($p<0.000$), satisfied with my life ($p<0.000$), optimistic about my future ($p<0.000$), life is worthwhile ($p<0.001$), connected to the people around me ($p<0.001$), interested in the world around me ($p<0.003$), overcome challenges ($p<0.009$), connected to the local environment ($p<0.014$), and physically active ($p<0.025$) (Appendix, Fig. 63). Visitors mean scores statistically significantly increased, but variability (SD) did not change for belonging to a neighbourhood ($p<0.010$), and variability increased for trust people ($p<0.010$) and happy with my life ($p<0.003$) (Appendix, Fig. 63). MWS results indicated that visiting museums and art galleries resulted in measurable and significant improvement in visitors' personal wellbeing, in the components of positive functioning (engagement, meaning, and purpose), satisfying life, self-esteem and resilience, vitality, and in social wellbeing, in the components of belonging and trust and supportive relationships.

Quantitative results indicated that visits to museums and art galleries resulted in positive, measurable and significant changes to visitor subjective wellbeing in all NAW indicators, including for both personal wellbeing and social wellbeing (Fig. 6.1, Michaelson et al. 2009). Visits measurably and statistically significantly improved visitors' personal wellbeing, in all the components, including: emotional wellbeing, with participants indicating an increase in positive feelings and reduction in negative feelings after visiting; positive functioning, as visitors felt more engaged, feeling that the visit had meaning and purpose and had increased their feelings of autonomy; vitality, with visitors noting they felt stronger and more physical active; self-esteem and reliance, with visitors indicating their self-esteem and ability to overcome challenges had improved and that life was worthwhile; life satisfaction improved as visitor experienced an increase in satisfaction. Visiting museums and art galleries measurably and significantly improved visitor social wellbeing, including the components of belonging, visitors' feeling of belonging to a local environment improved; and supportive relationships, visitors felt they had developed relationships and felt more connected to people after their visit. Quantitative results indicated that visits to museums and art galleries were supportive in the NEF five ways to wellbeing—to take notice, to be active, to connect, and to keep learning (Aked et al. 2008).

Qualitative results indicated that visitors to museums and galleries felt they had been principally supported in their personal wellbeing. A recurrent theme in visitor commentary at all museums and galleries was that visiting had supported people to learn, through enabling them to be inspired and excited to learn and interested to learn more. Visitor commentary indicated that learning was related to learning new information about a specific type of art, architecture, archaeology, or history. Occasionally learning was related to broader personal or national histories, and reinforcement of personal or national memories. Visitors to museums felt that their positive functioning had improved in the component of meaning and purpose. Visitor commentary indicated that their experiences at museums and galleries had been enjoyable and exciting, linked to helping them to take notice and to improve positive functioning in the component of engagement.

Thematic analysis of qualitative commentary indicated that participants felt their experiences had supported their personal wellbeing in the components of emotional wellbeing, specifically improving positive feelings and positive functioning, supporting participants to have high

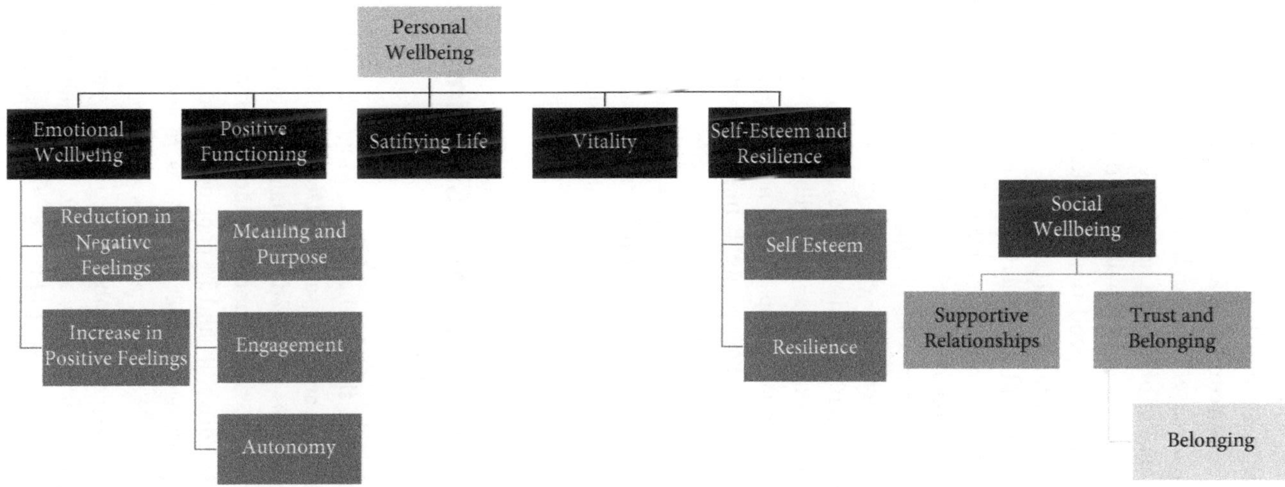

Fig. 6.1 Wellbeing indicators positively changed by visiting museums and galleries (Michaelson et al. 2009).

levels of engagement, and providing meaning and purpose. The results showed that visits to museums and galleries supported visitors to take notice and to keep learning. Thematic analysis of qualitative data indicated that visiting museums had an impact on social wellbeing, specifically developing a sense of belonging to a place. Comments reflected that a national sense of belonging had developed because of the material presented. Visitors occasionally noted museums and galleries had provided a space to spend time with family and friends and had improved supportive relationships.

Visits to museums and art galleries had a positive, measurable, and significant impact on visitors' subjective wellbeing in all NAW indicators, including both personal wellbeing and social wellbeing. Visiting museums and galleries had a significant and measurable positive impact on all components of personal wellbeing including positive functioning (engagement and meaning and purpose), self-esteem and resilience, vitality and satisfying life, and social wellbeing in both belonging and trust and supportive relationships. Visits had a measurable and significant impact on visitors' subjective wellbeing and provided visitor experiences that support the NEF five ways to wellbeing: visitors were helped to connect, take notice, keep learning, and to be active. The empirical evidence indicates that museums in this research had a measurable and significant positive impact on visitors' being supported in all categories and indicators except to give.

Heritage and Archaeological Sites' Impact on Visitor Wellbeing

Visitors to six heritage sites in the United Kingdom, United States, and India completed before and after surveys. Demographic analysis indicated that 73% of respondents identified as female, 26% as male, and 1% preferred not to answer; 95% of respondents indicated that they did not consider that they had a disability and 5% considered that they had a disability. In relation to age, 1% were under 20, 18% were aged 20–29, 12% were 30–39, 14% 40–49, 20% 50–59, and 35% were over 60. Demographic results indicated the typical respondent to the survey was female, did not have a disability, and was over the age of 60. Typical heritage and archaeological site respondents were of older age than museum and art gallery respondents.

After visiting heritage and archaeological sites, visitors' PANAS negative mean average scores decreased by 3.4% (Appendix, Fig. 64). PANAS negative visitor scores statistically significantly decreased and reduced in variability for irritable ($p<0.000$), distressed ($p<0.000$), upset ($p<0.001$), nervous ($p<0.002$), and afraid ($p<0.016$) (Appendix, Fig. 66). Visitor PANAS positive mean average scores increased by 1.9%, and scores statistically significantly increased and reduced in variability (SD) for inspired ($p<0.000$), determined ($p<0.001$), proud ($p<0.003$), active ($p<0.012$), enthusiastic ($p<0.012$), excited ($p<0.025$), interested ($p<0.036$), and attentive ($p<0.038$) (Appendix, Figs 65 and 66). Visitors' mean scores significantly increased and had increased variability for 'strong' ($p<0.001$) (Appendix, Fig. 66). PANAS results indicated that visiting heritage sites and archaeological sites measurably and significantly improved visitors' personal wellbeing in the components of emotional wellbeing (increase in positive feelings and reduced negative feeling), positive functioning (inspired, enthusiastic, excited, interested, attentive), vitality (active), and self-esteem and reliance (determined, proud).

MWS results indicated that visitor average mean scores across all measures increased by 1.1% after visiting heritage and archaeological sites (Appendix, Fig. 67). After visiting museums and galleries, visitor mean MWS scores statistically significantly increased and reduced in variability (SD) for feeling connected to the people around me ($p<0.000$), connected to the local environment ($p<0.014$), happy with my life ($p<0.014$), satisfied with my life ($p<0.021$), and I belong to a neighbourhood ($p<0.022$) (Appendix, Fig. 68). There was a statistically significant improvement and increase in variability (SD) in visitors feeling people support each other ($p<0.000$) and could trust people ($p<0.000$) (Appendix, Fig. 68). MWS results suggested that visiting heritage sites positively impacted on social wellbeing, in the components of supportive relationships (connected to people) and trust and belonging (connected to local environment), and personal wellbeing, in the components of satisfying life (satisfied with life) and emotional wellbeing (happy with life).

Quantitative results indicated that physical engagement in heritage and archaeological sites significantly improved visitors' ability to connect with people and their local environment, and improved their social wellbeing, helping them to develop supportive relationships and belonging to people and places. Visits positively affected visitors' personal wellbeing, and supported their emotional wellbeing in reducing negative feelings, increasing positive feelings, and improving visitors feeling happy with life. Heritage and archaeological sites improved visitors: positive functioning—they felt more engaged and that the visit had meaning and purpose; life satisfaction, linked to personal wellbeing component of satisfying life and supported participants to feel active; and vitality—they felt higher levels of self-esteem and reliance, feeling prouder and more determined (Fig. 6.2). Visits to heritage and archaeological sites measurably and significantly supported visitors to be connected, to be active, and to take notice, but not to keep learning or to give.

Thematic analysis of visitor commentary indicated that all case study heritage and archaeological site visits resulted in visitors feeling that their personal wellbeing had improved. In commentary, visitors noted positive changes to their emotional wellbeing and indicated that experiences at heritage and archaeological sites had supported them having positive feelings and emotions including happiness, calm, and peace whilst at the site. Visitors' commentary indicated that they felt their positive functioning had improved during the visit, that they felt engaged with the aesthetics and historical narratives surrounding the place and with the local environment. They indicated that their visit had meaning and purpose, and that they had learnt about art, architecture, and history. Visitors commented that heritage and archaeological sites had benefited their social wellbeing, by providing a place and space to meet friends and family to spend time together, and to develop existing supportive relationships with loved ones. Visitors noted that at some heritage and archaeology sites they had developed new relationships and connections with heritage professionals, staff, and volunteers and that the visit had supported their wider sense of belonging, linked to national, regional, or cultural identity.

Qualitative results indicated visitor experiences at heritage and archaeological sites positively impacted visitors' subjective wellbeing, as identified with NAW indicators of personal wellbeing and social wellbeing (Michaelson et al. 2009). Visitors felt that their personal wellbeing had improved in all indicators, including emotional wellbeing, positive functioning, vitality, satisfying life, self-esteem, and reliance. Visitors indicated that they felt their social wellbeing had improved, in the components of supportive relationships and belonging and trust. Results indicate that visiting heritage sites supported participants in four of the five NEF five ways to wellbeing, specifically to keep learning, connect, be active, and take notice. Visiting the heritage sites did not support measurably support visitors to give (Aked et al. 2008).

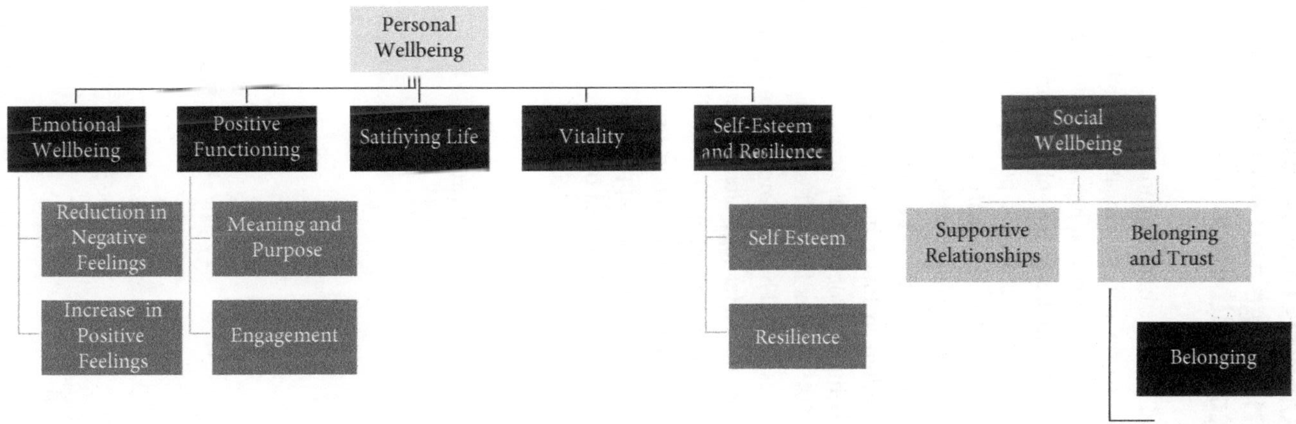

Fig. 6.2 National Accounts Framework wellbeing indicators improved by visiting heritage and archaeological sites (based on NAF, Michaelson et al. 2009).

Comparative Analysis of Museums and Art Galleries, and Heritage and Archaeological Sites' Impact on Visitor Wellbeing

Visiting museums and galleries had the most measurable and significant impact on visitor mean scores across all measures, including PANAS positive and negative and MWS (Appendix, Fig. 69). Visitors to museums and galleries demonstrated the highest level of measurable and significant improvements to all NAW indicators for personal wellbeing (emotional wellbeing, self-esteem and reliance, satisfying life, positive functioning, and vitality) and social wellbeing (supportive relationships and belonging) (Michaelson et al. 2009). Analysis indicated that visits to heritage and archaeological sites resulted in lower numbers of significant changes to visitors' personal wellbeing and social wellbeing.

Comparison of PANAS results from all case study sites indicated that museums and art galleries overall had the most optimal impact on visitors' hedonic wellbeing (Fig. 6.3). Heritage sites, specifically larger historic estates, had the lowest levels of impact, although still positive, on hedonic wellbeing (Fig. 6.3). This indicated that visitor practices at museums and art galleries resulted in visitors feeling happier, and having less negative feeling and more positive feeling, than that of visitors to heritage sites and archaeological sites.

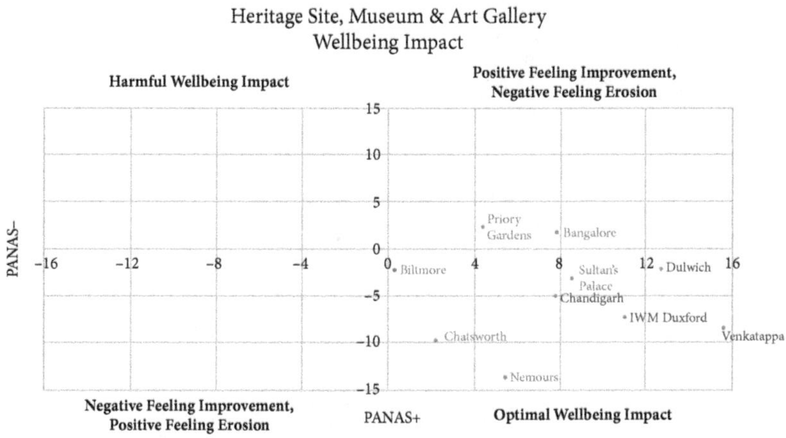

Fig. 6.3 Table indicating PANAS-linked hedonic wellbeing impact of heritage sites and archaeological sites.

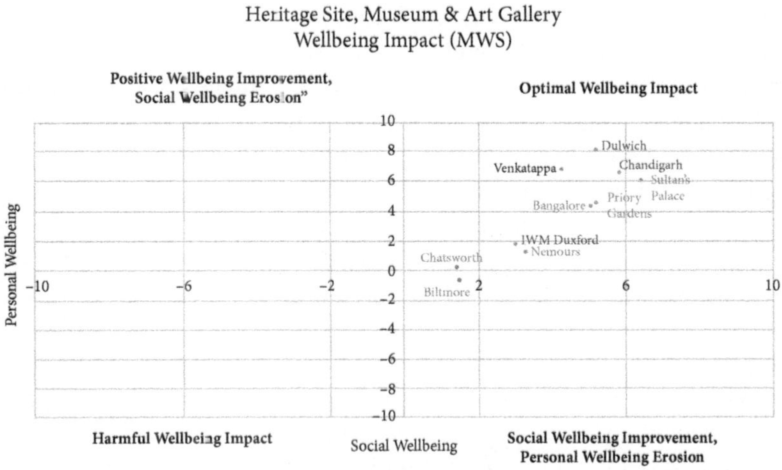

Fig. 6.4 MWS mean average score mapped out by personal and social wellbeing indicators for heritage sites, archaeological sites, museums, and galleries.

The categorization of MWS visitor mean average scores based on questions linked to social and personal wellbeing indicators (Fig. 6.4) enabled the comparison between all case study sites of their specific impact on visitors' personal and social wellbeing indicators. Results indicated TSSP had the highest impact on visitors' social wellbeing and DPG on visitors' personal wellbeing. TSSP, DPG, and CGMAG had the most optional impact on visitors' eudaimonic wellbeing (Fig. 6.4). Results indicated that museums and galleries had greatest impact on improving visitors' personal wellbeing and heritage and archaeological sites on improving visitors' social wellbeing (Fig. 6.4).

The research showed that both museums and galleries and heritage and archaeological sites did not measurably improve social wellbeing in the component of developing trust and did not support personal wellbeing, specifically improvement to feeling optimistic about the future linked to self-esteem and reliance. Heritage and archaeological sites did not support visitors to feel a sense of accomplishment linked to meaning and purpose and autonomy, and visitors did not report an improvement to freedom of choice and movement, both related to the indicator positive functioning.

Commentary after visits to heritage and archaeological sites suggested that there was a therapeutic benefit to visitors' personal wellbeing, specifically linked to emotional wellbeing and happiness, with words such as calm, relaxing, and peaceful frequently used. Visitors indicated therapeutic benefits and positive impacts of visits to heritage and archaeological sites on emotional wellbeing, which was directly linked to outdoor space and connection to nature. The therapeutic benefit of museums and arts galleries to visitors was not explicit in visitor commentary, indicating that, despite these places having a greater measurable impact on personal wellbeing including emotional wellbeing, visitors did not perceive museums and art galleries as places of therapeutic benefit, perhaps due to lack of outdoor provisions at museums and art galleries, and their physical experience usually being focused indoors.

Visits to heritage and archaeological sites resulted in negative impacts to emotional wellbeing, positive functioning, self-esteem, and social wellbeing linked to supportive relationships. Negative impacts to visitors' emotional wellbeing was linked to increases in negative feelings, including upset, guilty, and afraid, and visitors indicated potential reasons for these, which included: lack of freedom of choice, autonomy, and freedom of movement around sites linked to people feeling upset; high financial cost of visiting and ostentatious displays of wealth at some heritage sites, which resulted in some visitors feeling guilty; a reduction in feeling people supported each other, resulting in visitors feeling afraid and upset. A decrease in visitors feeling there was an opportunity to learn and feeling the visit was worthwhile could be related to visitors feeling more upset, as expectations did not relate to visitors' experiences, a 'jarring', and conflicts between personal expectation versus experience. These jarring moments at heritage and archaeological sites appear to be linked to poor preservation, conservation, and protection of aesthetics of site and issues with interpretation and lack of detail about histories and personal stories. Some negative impacts of visits to heritage assets on personal and social wellbeing could be linked to continued Covid restrictions, and the longer-term emotional impacts of Covid and social isolation.

Results indicated there were areas in which quantitative and qualitative data did not correlate, for example, despite heritage and archaeological sites having a measurable and significant impact on life satisfaction, visitors did not comment on this attribute. Perhaps this demonstrates the complexities of articulating and being consciously aware of wellbeing attributes.

Comparative analysis of visitor results from heritage and archaeological sites, museums, and art galleries suggested all heritage assets supported visitors to take notice, to connect, keep learning, and be active, but none of the sites appeared to support visitors to give. Results indicated that museums and art galleries had the most measurable and statistically significant impact on visitors in personal wellbeing and social wellbeing.

UK Heritage Places Impact on Wellbeing

Visitors to heritage places in the United Kingdom completed before and after surveys. Respondents' demographics indicated that 27% were over 60, 68% female, 97% white, and 91% did not have a disability. National demographic data from the United Kingdom indicated the UK population was 51% female and 49% male; and 22% were over 60, 27% aged 40–59, 29% aged 18–39; 86% white, and 22% disabled (ONS 2021). This indicated that respondents to the survey were skewed towards people who described their gender as female, white, aged over 60, and who did not consider themselves to have a disability.

After visiting UK heritage places, PANAS negative visitor mean average scores decreased by 7.3% and scores statistically significantly decreased and reduced in variability (SD) for feeling irritable, guilty, nervousness and distressed ($p<0.000$), and afraid ($p<0.016$) (Appendix, Fig. 70). After visiting UK heritage places, PANAS positive mean average visitor scores increased by 6.8% (Appendix, Fig. 69) and scores statistically significantly increased and reduced in variability (SD) for feeling proud, active, determined ($p<0.000$), attentive ($p<0.002$), enthusiastic ($p<0.006$), and interested ($p<0.028$) (Appendix, Fig 70). MWS visitor average mean scores increased by 3.7% after visiting UK heritage places and scores statistically significantly increased and reduced in variability (SD) for satisfied with my life ($p<0.004$), sense of accomplishment ($p<0.000$), people support each other ($p<0.030$), overcome challenges ($p<0.042$), and optimistic about my future ($p<0.048$) (Appendix, Fig. 69, Fig. 71). Visits to UK heritage places measurably and statistically significantly improved visitors' personal wellbeing, in the components of emotional wellbeing as visitor negative emotions were reduced; positive functioning, specifically engagement and meaning and purpose (interest, enthusiasm, accomplishment, attentive, determined); vitality, with visitors feeling more active; life satisfaction and resilience and self-esteem; and in social wellbeing, in building supportive relationships.

Gender of Visitors at UK Heritage Places

After visiting UK heritage places, female visitors' PANAS negative average mean scores decreased by 7.6%, and statistically significantly decreased and reduced in variability for feeling irritable, distressed, nervous ($p<0.000$), jittery ($p<0.001$), guilty ($p<0.021$), scared ($p<0.040$), and afraid ($p<0.045$) (Appendix, Fig. 72). Female visitors' PANAS positive mean average scores increased by 6.8%, and scores statistically significantly improved and reduced in variability for feeling determined and proud ($p<0.001$) (Appendix, Fig. 72). Female visitors' MWS mean average scores increased by 2.3% and scores statistically significantly improved and reduced in variability for sense of accomplishment ($p<0.000$), interested in world around me ($p<0.041$), optimistic about the future ($p<0.027$), and people support each other ($p<0.027$); there was a significant improvement but no change to variability in: satisfied with life ($p<0.003$) and trust people ($p<0.007$) (Appendix, Fig. 72).

PANAS and MWS results indicated that visits to UK heritage places measurably and significantly improved female visitors' personal wellbeing, in the components of emotional wellbeing and a reduction in negative feelings, and an increase in positive feeling; positive functioning, increasing meaning and purpose; and engagement, and self-esteem and resilience as they felt prouder, determined, and more optimistic; and satisfying life. Visits supported social wellbeing in the components of trust and belonging and supportive relationships. Visiting UK heritage places supported female visitors to connect and to take notice.

After visiting UK heritage places, male visitors' PANAS negative mean average scores reduced by 6.2%, and scores statistically significantly decreased and reduced in variability for feeling irritable ($p<0.000$), distressed ($p<0.041$), and nervous ($p<0.041$) (Appendix, Fig. 72). Male visitors' PANAS positive mean average scores increased by 6.1% and scores statistically significantly increased and reduced in variability for feeling proud ($p<0.003$) and attentive ($p<0.033$) (Appendix, Fig. 72). Male visitors' MWS mean average scores increased by 0.9% after their visits (Appendix, Fig. 72). PANAS and MWS results indicated that visits to UK heritage places measurably and significantly improved male visitors' personal wellbeing, in the components of; emotional wellbeing, reducing negative feelings; positive functioning, improving engagement; and self-esteem and resilience. Visiting UK heritage places supported male visitors to take notice of the world around them.

The gender of visitors made a measurable difference to the impacts that UK heritage sites had on visitors' personal and social wellbeing. After visiting UK heritage places, female visitors' personal wellbeing improved in the categories of positive functioning, satisfying life, and social wellbeing improved in the components of trust and belonging and supportive relationships. Male visitors' personal wellbeing improved in the component of positive functioning and there was no significant change to social wellbeing. UK heritage places improved female visitors' personal wellbeing and social wellbeing, but for male visitors there was only an improvement to their personal wellbeing. There was a more significant and measurable impact on the personal and social wellbeing of female visitors compared with males.

Age of Visitors at UK Heritage Places

After visiting UK heritage places, visitors under 30 years of age showed PANAS negative mean average scores reduced by 11.6%, and scores statistically significantly decreased and reduced in variability for distressed ($p<0.004$), nervous ($p<0.022$), and irritable ($p<0.031$) (Appendix, Fig. 73). The PANAS positive mean average scores for visitors under 30 years old increased by 9.4%, and MWS mean scores increased by average of 3.3%, with statistically significant increases and reduced variability in interested ($p<0.007$) and active ($p<0.015$) (Appendix, Fig. 73). PANAS and MWS scores indicated visiting UK heritage places measurably and significantly improved the personal wellbeing of visitors under 30, in the components of emotional wellbeing, positive functioning (engagement), and vitality, and supported them to be active and take notice of the world around them.

After visiting UK heritage places, the PANAS negative mean average scores of visitors aged 30–39 reduced by 5.4%, and scores statistically significantly decreased and reduced variability for feeling irritability ($p<0.004$) and nervousness ($p<0.027$), and PANAS positive mean average scores increased by 1.8% (Appendix, Fig. 73). Their MWS average mean scores increased by 4.0%, and scores statistically significantly increased and reduced variability for feeling strong ($p<0.008$), proud ($p<0.043$), enthusiastic ($p<0.043$), and sense of accomplishment ($p<0.049$) (Appendix, Fig. 73). PANAS and MWS results indicated that visits to UK heritage places measurably and statistically significantly improved the personal wellbeing of visitors aged 30–39, specifically regarding emotional wellbeing, self-esteem and resilience, and positive functioning, and supported them to take notice.

After visiting UK heritage places, the PANAS negative mean average scores of visitors aged 40–49 reduced by 11.6%, and scores statistically significantly decreased and reduced in variability for feeling irritable ($p<0.001$), afraid ($p<0.021$), nervous ($p<0.039$), and guilty ($p<0.039$), and PANAS positive mean average scores increased by 9.4% (Appendix, Fig. 73). Their MWS mean average scores increased by 3.3%, and scores statistically significantly increased and reduced in variability for accomplishment ($p<0.000$), satisfied with life ($p<0.021$), inspired ($p<0.003$), active ($p<0.023$), and determined ($p<0.041$) (Appendix, Fig. 73). PANAS and MWS results indicated that visits to UK heritage places measurably and statistically significantly improved the personal wellbeing of visitors aged 40–49, in the components of emotional wellbeing, positive functioning, vitality, and satisfying life, and supported them to take notice of the world around them and be active.

After visiting UK heritage places, the PANAS negative mean average scores of visitors aged 50–59 decreased by 5.4%, and scores statistically significantly decreased and reduced in variability for feeling irritable ($p<0.000$) and distressed ($p<0.021$), and PANAS positive mean average scores increased by 4.5% (Appendix, Fig. 73). Their MWS mean average scores increased by 0.1%, and scores statistically significantly increased and reduced in variability for feeling proud ($p<0.003$) (Appendix, Fig. 73). PANAS and MWS results indicated that visits to UK heritage places measurably and statistically significantly improved the personal wellbeing of visitors aged 50–59, in the components of emotional wellbeing and self-esteem, and supported them to take notice.

After visiting UK heritage places, the PANAS negative mean average scores of visitors aged 60 and over reduced by 5.6%, and scores statistically significantly decreased and reduced in variability for feeling nervous ($p<0.002$) and irritable ($p<0.023$) (Appendix, Fig. 73). Their PANAS positive mean average scores increased by 7.7% and scores statistically significantly increased and reduced variability for feeling inspired ($p<0.001$), proud ($p<0.004$), and enthusiastic ($p<0.031$) (Appendix, Fig. 73). Their MWS mean average scores increased by 2.3% and scores statistically significantly increased and reduced in variability for the opportunity to learn ($p<0.022$) (Appendix, Fig. 72). Quantitative results indicated that visits to UK heritage places measurably and significantly improved the personal wellbeing of visitors aged over 60, in the components of emotional wellbeing, positive functioning (mean and purpose, and engagement), and self-esteem, and supported them to keep learning and to take notice of the world around them.

All age groups visiting UK heritage places showed a measurable improvement to personal wellbeing and emotional wellbeing, including a reduction in negative feelings, specifically irritably, and improvement in positive feelings. Visits had the most significant and measurable impact on the personal wellbeing of visitors under 30 and aged 40–49, and the least measurable and significant impact on visitors aged 30–39 and 50–59. Except for visitors aged 50–59, there was a statistically significant improvement within all age groups to positive functioning, and improvement in taking notice. No age groups indicated any improvement to social wellbeing. Specific impact to visitors' personal wellbeing was dependent on the age of participants; for example, there was a significant improvement to self-esteem and reliance of visitors aged 30–39 and 50–59, and a significant improvement to vitality for visitors aged under 30. Visitors aged 40–49 were supported to be active, and visitors aged 60 and over were supported to keep learning.

Summary UK Heritage Places

Visits to UK heritage places significantly improved visitors' personal wellbeing, in the components of positive functioning, vitality, life satisfaction and reliance, and self-esteem, and their social wellbeing, in the component of supportive relationships. Visitor experiences at UK heritage places supported individuals to connect, to be active, and to take notice of the world around them. Results indicated that gender affected the impact on wellbeing; only female visitors' social wellbeing was measurably improved, and despite a statistically significant improvement in the personal wellbeing of both male and female visitors, male visitors' improvement to personal wellbeing was measurably lower (Appendix, Fig. 72). UK heritage sites supported both male and female visitors to take notice, but only female visitors to connect to the world and to people around them. There were measurable differences in the wellbeing impact of visits to UK heritage places based on the age and gender of visitors and indicated the potential role of heritage visits in reducing wellbeing inequality, and supporting inclusivity and diversity, including the ability to support visitors under the age 30 to take notice and to be active, and to support visitors aged over 60 to keep learning and take notice.

US Heritage Places Impact on Visitor Wellbeing

Visitors completed before and after surveys from US heritage places. Respondent demographics indicated that the plurality were over 60 (46%), and the majority described their gender as female (79%), described their background as white (93%), and did not consider themselves to have a disability (95%). US demographics indicated 51% female, 72% white, 17% over 65, and 9% disabled (United States Census Bureau 2020). This indicated that the respondent demographic was skewed towards individuals who describe their gender as female, ethnicity as white, age as over 60, and non-disabled.

Visitors to US heritage places experienced average reduction of 3.0% in PANAS negative mean scores after visiting, with a statistically significant decrease and reduced variability in feeling irritable ($p<0.000$), distressed ($p<0.000$), upset ($p<0.004$), nervous ($p<0.016$), and afraid ($p<0.029$) (Appendix, Fig. 74). Visiting US heritage places increased visitor mean average PANAS positive scores by 0.6%, and statistically significantly increased and reduced variability in feeling inspired ($p<0.014$) (Appendix, Fig. 74). After visiting US heritage sites, MWS visitor mean scores increased by an average of 0.2%, and there was a statistically significant increase and reduced variability in feeling connected to the people around me ($p<0.006$) (Appendix, Fig. 75). MWS scores also indicated that visitors to US heritage places also experienced a statistically significant decrease and reduced variability in feeling they had freedom of choice ($p<0.029$) and could overcome challenges ($p<0.047$), and a statistically significant reduction and increased variability for opportunity to learn new things ($p<0.004$) (Appendix, Fig. 75). Results suggested that visiting US heritage places measurably and significantly improved personal wellbeing, in the component of emotional wellbeing, and social wellbeing, in the component of supportive relationships. Results indicated that visits to US heritage places also had a negative impact on visitors' personal wellbeing in the categories of positive functioning, in the components of autonomy and meaning and purpose, and in self-esteem and resilience. This indicated that visitors to US heritage places felt supported to connect, but did not feel supported to keep learning, take notice, be active, or to give.

Gender at US Heritage Places

Female visitors to US heritage sites experienced a reduction of –3.1% in mean PANAS negative scores after visiting, and statistically significant decreases and reduced variability in feeling distressed ($p<0.000$), irritable ($p<0.000$), and nervous ($p<0.015$); and their PANAS positive mean scores increased by 3.17% (Appendix, Fig. 76). Female visitors' mean average MWS scores increased by 0.04% and they experienced a statistically significant increase and reduced variability in 'connected to the people around me' ($p<0.007$) (Appendix, Fig. 76). Female visitors also experienced statistically significant reductions and reduced variability in feeling afraid ($p<0.033$) and having freedom of choice ($p<0.028$). Results indicated US heritage sites measurably and significantly improved female visitors' personal wellbeing, in the component of emotional wellbeing, and social wellbeing, in the component of supportive relationships. Results also demonstrate that visiting US heritage sites had a significant and measurable negative impact on female visitors' positive functioning, specifically on their autonomy.

Male visitors to US heritage places experienced a reduction of –2.8% in PANAS negative average mean scores after visiting, a statistically significant decrease, and reduced in variability for feeling distressed ($p<0.004$); PANAS positive scores increased by 1.7% and there was a statistically significant improvement and reduced variability for feeling excited ($p<0.015$), inspired ($p<0.030$), and determined ($p<0.034$) (Appendix, Fig. 76). Male visitors' mean average MWS score increased by an average of 0.2% (Appendix, Fig. 76). These results indicated that after visiting US heritage sites male visitors experienced an improvement to personal wellbeing, in the component of positive functioning, specifically to engagement and emotional wellbeing, a reduction of negative feelings and an increase in positive feelings.

Results indicated that the gender of visitors impacted the components of wellbeing. Whilst both female and male visitors experienced statistically significant improvements to personal wellbeing, specifically to emotional wellbeing, only female visitors experienced improvements to social wellbeing (in the component of supportive relationships), and only male visitors experienced improvement to personal wellbeing in the component of positive functioning (engagement). Visits had a negative impact on female visitors' personal wellbeing, in the component of positive functioning (autonomy), and emotional wellbeing, linked to feeling more afraid. Through visiting heritage places, female visitors were supported to connect and male visitors to take notice.

Disability at US Heritage Places

The United States was the only country that provided sufficient data to undertake comparative statistical analysis of disabled and non-disabled visitors to heritage sites.

After visiting US heritage places, disabled visitors PANAS negative average mean scores reduced by –9.1% and scores statistically significantly decreased and reduced in variability for distressed ($p<0.004$) (Appendix, Fig. 77). Their PANAS positive average mean scores increased by 2.5%, and MWS mean average scores increased by 2.0% (Appendix, Fig. 77). Visits positively improved disabled visitors' personal wellbeing, linked to improvement in emotional wellbeing, specifically a reduction in negative feelings.

After visiting US heritage places, non-disabled visitors' PANAS negative mean average scores decreased by –2.7%, and scores statistically significantly reduced and decreased in variability for feeling distressed, irritable ($p<0.000$), afraid ($p<0.027$), and nervous ($p<0.038$) (Appendix, Fig. 77). Non-disabled visitors' PANAS positive average mean scores increased by 0.5% and MWS mean average scores decreased by 0.05%, with a statistically significant increases and reduced variability in 'I feel connected to people around me' ($p<0.006$) (Appendix, Fig. 77). This indicated that visiting heritage sites positively improved non-disabled visitors' personal wellbeing, in the component of emotional wellbeing (reduction in negative feelings), and social wellbeing, in the component of supportive relationships.

Results indicated that both disabled and non-disabled visitors experienced statistically significant improvements to personal wellbeing, specifically in the component of emotional wellbeing, linked to a reduction in negative feeling, specifically to distressed. Visits had the greatest measurable impact on the personal wellbeing of disabled visitors (Appendix, Fig. 77). Yet non-disabled visitors to US heritage place experienced a greater number of statistically significant changes to wellbeing indicators, including personal wellbeing, specifically emotional wellbeing, and social wellbeing, specifically supportive relationships. Visiting US heritage place supported non-disabled people, but not disabled people, to connect.

Age at US Heritage Places

After visiting US heritage places, the PANAS negative mean average scores of visitors under 30 years old reduced by –2.6%, PANAS positive mean average scores increased by 1.1% and MWS mean average scores decreased by –4.1% (Appendix, Fig. 78). Visiting US heritage places measurability improved emotional wellbeing, but overall had a negative impact on personal and social wellbeing of visitors under 30.

After visiting US heritage places, the PANAS negative mean average scores of visitors aged 30–39 reduced by –4.7%; scores statistically significantly decreased and reduced in variability for feeling upset ($p<0.004$) (Appendix, Fig. 78), and PANAS positive mean average scores increased by 1.2%. MWS mean average scores for visitors aged 30–39 increased by 1.6% (Appendix, Fig. 78). Visits measurably and significantly improved the personal wellbeing of visitors aged 30–39, in the category of emotional wellbeing.

After visiting US heritage places, the PANAS negative mean average scores of visitors aged 40–49 reduced by –4.3% and scores statistically significantly decreased and reduced in variability for feeling irritability ($p<0.007$) and distressed ($p<0.012$) (Appendix, Fig. 78). The PANAS positive mean average scores of visitors aged 40–49 increased by 1.5%, their MWS mean average scores increased by 0.2%, and scores statistically significantly increased and reduced in variability for 'feel physically active' ($p<0.029$) (Appendix, Fig. 78). Visits measurably and significantly improved the personal wellbeing of visitors aged 40–49, in the categories of emotional wellbeing and vitality, and supported them to be active.

The PANAS negative mean average scores of visitors aged 50–59 to US heritage places decreased by 3.8%, and scores statistically significantly decreased and reduced in variability for feeling distressed ($p<0.004$) (Appendix, Fig. 78). The PANAS positive mean average scores of visitors aged 50–59 increased by 2.0%, and MWS mean average scores increased by 0.7% (Appendix, Fig. 78). Visiting US heritage places measurably and statistically significantly improved the personal wellbeing of visitors aged 50–59, specifically their emotional wellbeing.

The average PANAS negative mean average scores of visitors over the age of 60 to US heritage places reduced by –2.2%, and their scores statistically significantly decreased and reduced in variability for feeling irritable ($p<0.001$) and distressed ($p<0.007$) (Appendix, Fig. 78), and their PANAS positive mean average scores increased by 0.6%. MWS mean scores for visitors aged 30–39 increased a by an average of 0.2%, and scores statistically significantly increased and reduced variability for 'I can trust people' ($p<0.028$) (Appendix, Fig. 78). Visits measurably and significantly improved the personal wellbeing of visitors aged over 60, specifically their emotional wellbeing, and social wellbeing, linked to improved trust and belonging, and supported them to connect.

Comparison of quantitative results indicated specific trends in how visits to US heritage sites impacted the personal and social wellbeing of different visitor age groups. Irrespective of visitor age, US heritage places statistically significantly improved visitors' personal wellbeing, in the component of emotional wellbeing, by reducing negative feelings. For example, after visiting US heritage sites people aged 40 and over felt significantly less distressed. People aged 30–39 experienced the largest improvement to wellbeing, as indicated by positive changes to wellbeing scores. People under the age of 30 overall experienced a negative change in their personal and social wellbeing. After visiting US heritage places, people aged 40–49 were the only age group that had a statistically significant improvement to personal wellbeing in the component of vitality and supported to be active. After visiting US heritage places, visitors over 60 were the only group that experienced a statistically significant improvement to their social wellbeing, specifically in trust.

Summary US Heritage Places

Quantitative results indicated that visiting US heritage places had a significant and measurable impact on visitors' personal wellbeing and social wellbeing. After their visits, their personal wellbeing improved; they experienced significant reductions to negative feelings and increases to positive feelings, indicating an improvement to emotional wellbeing and higher levels of engagement linked to an improvement in positive functioning. Visitors indicated that their experience at US heritage sites had a significant positive impact on social wellbeing, specifically on developing supportive relationships. The greatest positive impact was to the personal wellbeing of non-disabled female visitors aged 30–39, and the most significant and measurable improvement to social wellbeing was to female non-disabled visitors over 60. Yet heritage places in the United States had a negative measurable impact on the personal and social wellbeing of visitors under 30. Overall, visits to US heritage sites supported visitors to take notice of the world around them and to connect with people and places.

Indian Heritage Places Impact on Visitor Wellbeing

Visitors to Indian heritage places completed before and after surveys. Respondent demographics indicated most were aged 20–29 (74%), described their gender as female (52%), and did not consider themselves to have a disability (99%). Indian national demographic data indicates 48% of population are female and 52% male, suggesting slightly higher numbers of female respondents in comparison to the national average (World Bank Group 2023). Comparison of age and disability data was not possible, due to large age ranges, 25–54, and because the India national census does not collect disability data (World Bank Group 2023). Data relating to background and ethnicity was not collected at Indian heritage places, due to sensitivity concerning religious and caste identity. Analysis of demographic data in India was complicated by high levels of visitors aged 30 and under, and interestingly no visitors indicated they had a disability. For this reason, a statistical analysis of impacts of heritage places on specific visitor demographics was not undertaken for India.

PANAS negative mean average negative scores for visitors to Indian heritage places decreased by 3.6% and scores statistically significantly decreased and reduced in variability (SD) for feeling scared ($p<0.003$) and upset (<0.006), and significantly decreased but with unchanged variability for feeling irritable ($p<0.005$) (Appendix, Fig. 79). Visitors' PANAS positive mean scores indicated an average increase by 9.7% and scores statistically significantly increased and reduced in variability (SD) for excited ($p<0.000$) and active ($p<0.000$) (Appendix, Fig. 79). Visitor MWS average mean scores increased by 5.7% and scores statistically significantly increased and reduced in variability (SD) for feeling able to overcome challenges ($p<0.000$), optimistic about my future ($p<0.000$), satisfied with my life ($p<0.000$), connected to the people around me ($p<0.000$), and interested in the world around me ($p<0.000$) (Appendix, Fig. 80). Visits to

Indian heritage places measurably and statistically significantly improved personal wellbeing, in the components of emotional wellbeing, as visitors' negative emotions were reduced; positive functioning, specifically improved engagement (excited) and vitality with visitors feeling more active; life satisfaction, and resilience and self-esteem; and in social wellbeing, in the component of building supportive relationships.

Summary Indian Heritage Places Impact on Visitor Wellbeing

Indian heritage places had a statistically significant impact on visitors' personal wellbeing and social wellbeing. Visitors' personal wellbeing measurably improved in the categories of emotional wellbeing, positive functioning, life satisfaction, vitality, and self-esteem and reliance, and visitors' social wellbeing improved in the category of supported relationships. Indian heritage places supported visitors to connect to people around them, to take notice of the world around them, and to be active.

Comparison of US, UK, and Indian Heritage Places' Impact on Visitor Wellbeing

Visiting heritage places in all the countries in this study (the United States, the United Kingdom, and India) resulted in a significant and measurable positive change to visitors' social wellbeing and personal wellbeing. Heritage places in all countries had a positive impact on visitors' personal wellbeing (in the components of emotional wellbeing and positive functioning), and on their social wellbeing, in the components of supportive relationships. However, comparative analysis of quantitative and qualitative results from visitors' surveys at heritage places in the United States, the United Kingdom, and India, showed that there were 'national' differences in the impact these places have on visitor wellbeing.

Indian heritage places had the most measurable and significant impact on visitors' subjective wellbeing, on both personal wellbeing and social wellbeing. Visitors to Indian heritage places experienced the highest number of statistically significant increases to mean scores, yet the highest levels of variability in results in comparison to the United Kingdom and United States. Indian heritage places improved visitors' personal wellbeing components including emotional wellbeing, vitality, positive functioning, self-esteem and resilience, life satisfaction and to social wellbeing, and supportive relationships. Indian heritage places supported visitors to connect, to keep learning, to take notice, and to be active. Qualitative data indicated that, in comparison to the United Kingdom and United States, Indian heritage places received the highest level of negative feeling and emotions. Elevated levels of negative emotions and feelings were associated with concerns about the presentation of artefacts and the conservation of the sites, and in specific cases political ideologies that specific heritage management choices were seen to reflect, for example deliberately allowing Mughal heritage to decay. Despite elevated levels of negative emotions in visitor commentary, these did not have a negative quantifiable impact on personal or social wellbeing. This might suggest that negative emotions did not, in the case of Indian heritage places, have a negative impact on visitors' wellbeing.

UK heritage places had the greatest quantitative positive impact to visitors' personal wellbeing, linked to an improvement of all the subcomponents including emotional wellbeing, life satisfaction, resilience and self-esteem, vitality, positive functioning (meaning and purpose and engagement), and impacted visitors social wellbeing in the component of supportive relationships. Visitor age was a significant factor in the impact of UK heritage places on personal wellbeing. Qualitative data indicated a dominance within visitor commentary of the impacts of UK heritage sites on visitor social wellbeing, specifically to their sense of belonging and trust and development of supportive relationships. Visitors reported that UK heritage sites supported their personal wellbeing, linked to emotional wellbeing and positive functioning, including to learn, and to get outdoors and be physically active, and improved their vitality. Visitor commentary at UK heritage sites explicitly noted positive changes to wellbeing, and included comments such as calm and peaceful. This suggests that visitors to UK heritage places might be more aware of the changes that visiting heritage places could and did have to their wellbeing and mental health.

In comparison to UK and Indian heritage places, US heritage sites had the most significant and measurable positive impact on visitors' social wellbeing, in the component of supportive relationships. Corresponding qualitative data from visitor commentary indicated that visitors' experiences at US heritage places were driven and centred around leisure activities and around connecting with friends, family, and places, and benefited their social wellbeing. Whilst visitors to US heritage places experienced positive changes to emotional wellbeing and positive functioning, US heritage sites had the lowest comparable impact to visitor subjective wellbeing, specifically personal wellbeing, of any country in this study. US heritage places were unique in having measurable and significant negative impacts to personal wellbeing, including a decrease in visitors' self-esteem and positive functioning, linked to low levels of autonomy, and meaning and purpose. This indicated that visitors to US heritage places did not feel they had choice, freedom of movement, or a meaningful experience.

Heritage places in the United Kingdom, United States, and India did not have a statistically significant impact on visitors' learning, except for visitors to UK heritage sites over the age of 60. However, qualitative data from visitor commentary indicated that large numbers of visitors felt that they had 'learnt something new'. Visitors to UK and Indian heritage places had a statistically significant improvement to their vitality and felt supported to be active; visitor commentary linked this to being outside and in the fresh air. The measurable and significant improvement to feeling active and vitality was, in the case of the United Kingdom, associated with visitors aged 30–39 and 40–59. The vitality of visitors to US sites did not statistically significantly improve, which may be connected to visitors to US sites being the group transported by shuttle buses around the grounds.

Results indicated that age and gender of visitors influenced the impact that US and UK heritage places had on their personal and social wellbeing. UK and US heritage places had a specific and measurable impact on the social and personal wellbeing on male and female visitors, with differences between male and female visitors' wellbeing most measurable in the United Kingdom. Visits to heritage places in both the United Kingdom and United States statistically significantly improved the personal wellbeing of both male and female visitors, in the components of emotional wellbeing and positive functioning, and supported them to 'take notice'. After visiting US and UK heritage sites, female visitors were the only gender that indicated significant improvement to their vitality; they also felt supported to be active after visiting UK heritage places. Female visitors to US heritage places experienced significant negative changes to emotional wellbeing (feeling more afraid), and to positive functioning, feeling they had less autonomy. Female visitors to heritage sites in the United Kingdom and the United States experienced a statistically significant improvement in social wellbeing, in the component of supportive relationships, and in the United Kingdom also to belonging and trust. Results indicated that visits to heritage places had the greatest impact (both positive and negative) on female visitors' personal and social wellbeing.

Visitors of all age groups to UK and US heritage places experienced positive changes to their personal wellbeing, in the component of emotional wellbeing. The impact of visits on wellbeing varied with both the age of the visitor and the country where the site was located. For example, after visiting UK and US heritage places the vitality of people aged 40–49 significantly improved, and they felt more active. Results indicated country-specific age-based changes to wellbeing: for example, visits to heritage sites had the greatest measurable change to visitor wellbeing in the United Kingdom for people aged 40–49 and those under 30, and in the United States to people aged 30–39. Conversely, visitors to heritage sites that experienced the least measurable and elevated level of negative changes to wellbeing were those under 30 in the United States, and those aged 30–39 and 50–59 in the United Kingdom. These results suggest that the age groups that experienced the greatest positive change to their wellbeing because of visiting heritage sites in the United Kingdom experienced the least change to wellbeing in the United States, and vice versa. This potentially indicates how different national trends in visitor engagement and experience strategies at heritage sites impact different age groups. For example, visitor experiences at UK heritage places measurably improved the self-esteem and reliance of visitors aged 50–59 but did not measurably improve the positive functioning of visitors aged 30–39 and 50–59 and did not support them to take notice. Visitors aged over 60 experienced a significant impact to personal and social wellbeing after visiting UK and US heritage places; in the United States it improved their social wellbeing, specifically trust and belonging, and supported them to connect. In the United Kingdom, it improved their positive functioning and supported them to keep learning. After visiting UK heritage places, all age groups experienced higher levels of positive changes to personal wellbeing than visitors to US heritage sites. Whilst visiting UK heritage places did not measurably improve the social wellbeing of any age group, the social wellbeing of those over 60 showed an improvement after visits in the United States.

Visitors to all heritage places experienced a significant quantitative and qualitative improvement in their social wellbeing, through being supported to connect to family and friends, and by providing space, place, and time to develop supportive relationships. Yet, age, gender, and disability of the visitor significant influenced heritage places' ability to improve visitors' social wellbeing. Only visitors identifying as non-disabled, female, or over 60 experienced significant improvements to social wellbeing. Except for visitors over the age of 60, all other visitors to heritage places in the United Kingdom, India, and United States did not experience a statistically significant improvement in trust and belonging.

Data indicated that there were national differences in visitor motivations to visit heritage places, and these differences had an impact on visitor wellbeing. For example, in the United States the primary motivation for visits to heritage sites was for leisure and entertainment to improve social wellbeing; in the UK, visitors' motivation to visit was driven by improvement to personal wellbeing, specifically to emotional wellbeing and positive functioning, and often linked to informal learning. In India, visitors were motivated by improvement to personal wellbeing and positive functioning, often linked to formal learning, and the meaning and relevance of the heritage place to national identity.

Conclusion

This chapter has demonstrated both quantitatively and qualitatively that the impact on visitors' personal wellbeing of visits varies both by type of heritage site and by country. These results highlight the effect of different national approaches to heritage management including visitor engagement strategies, and how the perceptions and expectations of the visitor to the value of heritage places impacts the benefits to visitor wellbeing. Yet, despite place, and socio-cultural and political context, all heritage places in this study improved visitors' personal and social wellbeing and provided visitors with experiences that enabled them to connect with people and places and take notice of the world around them.

Heritage and Wellbeing: The Impact of Heritage Places on Visitors' Wellbeing. Faye Sayer, Oxford University Press.
 DOI: 10.1093/9780191914539.003.0006

7

Healthier Societies through Heritage

Heritage, Happiness, and Human Rights

This book has sought to provide internationally applicable evidence of the impact of heritage places on visitors' subjective wellbeing. The research described in this book has provided empirical qualitative and quantitative evidence of the impact on visitor wellbeing of heritage places. This research used a range of tested and newly developed measures to assess changes to individuals' subjective hedonic and eudaimonic wellbeing; this including measuring change to specific wellbeing attributes and capabilities after visiting heritage places.

The breadth and robust nature of this research has enabled this chapter to present guidelines and strategies for future best practice for how heritage places and projects can best support visitors' personal and social wellbeing, and how they can develop programmes and projects specifically tailored to support wider community wellbeing. At this critical juncture in wellbeing policy and practice, the research presented in this book has demonstrated that the heritage sector has value to society and can support global societal goals such as reducing wellbeing inequality and creating healthier societies. Research has demonstrated that by applying best practice models, based on evidence outlined in this book, heritage can contribute positively to individual and societal wellbeing and wider mental health. The preceding chapters have provided evidence to enable heritage organizations and institutions to recognize the impact of heritage on both social and personal wellbeing, and to leverage this evidence to help position heritage places in wider government wellbeing and health policy practice and agendas.

Thematic Pathways to Wellbeing

This book has shown that visiting heritage places including museums, art galleries, heritage sites and archaeological sites has a measurable and significant impact on the personal and social wellbeing of visitors (Fig. 7.1). Visiting heritage places had a significant impact on individuals hedonic and eudaimonic wellbeing, increasing positive feelings, happiness and life satisfaction (hedonic), and improving capabilities linked to positive functioning, vitality, self-esteem and resilience, belonging, trust and relationships.

Heritage places have significant impact on the key measurement criteria for personal wellbeing, including:

- *Quality of Life*: Visitors reported measurable and significant improvement to personal wellbeing; this included improvement to visitors' vitality and life satisfaction after visiting heritage places (Mathews and Izquierdo 2009).
- *Life Satisfaction*: There was a measurable and significant improvement to visitor self-esteem after visiting heritage sites (Knight et al. 2014: 11, Graham 2011).
- *Subjective Wellbeing*: Visitors reported a measurable and significant improvement in personal and social wellbeing, including emotional wellbeing and positive functioning, after visiting heritage places.
- *Happiness*: Visitors' feeling of contentment and pleasure improved after visiting heritage places, and their emotional wellbeing improved (Graham 2011).

This research demonstrated that heritage sites had a significant impact on the social wellbeing of visitors, improving visitors' feelings of connection to people and places and sense of belonging to areas and places, and trust in people. This indicated the wellbeing value of heritage is not only linked to hedonic and personal wellbeing, including quality of life, life satisfaction, and happiness, but also to social wellbeing and wider eudaimonic wellbeing capabilities, which include positive functioning, self-esteem and resilience, belonging and trust, and supportive relationships.

Through the management, interpretation, presentation, and conservation of the past, heritage places provide visitors with experiences that positively impact visitors' subjective wellbeing, including their personal and social wellbeing. Through management approaches and visitor engagement strategies, heritage places in this study supported visitors' wellbeing capabilities and functions. This research demonstrates, as previously asserted by Hodder, that heritage places' ability to positively impact individual subjective wellbeing capabilities and functions reflects their wider ability to support human rights (Hodder 2010). As a result of improving

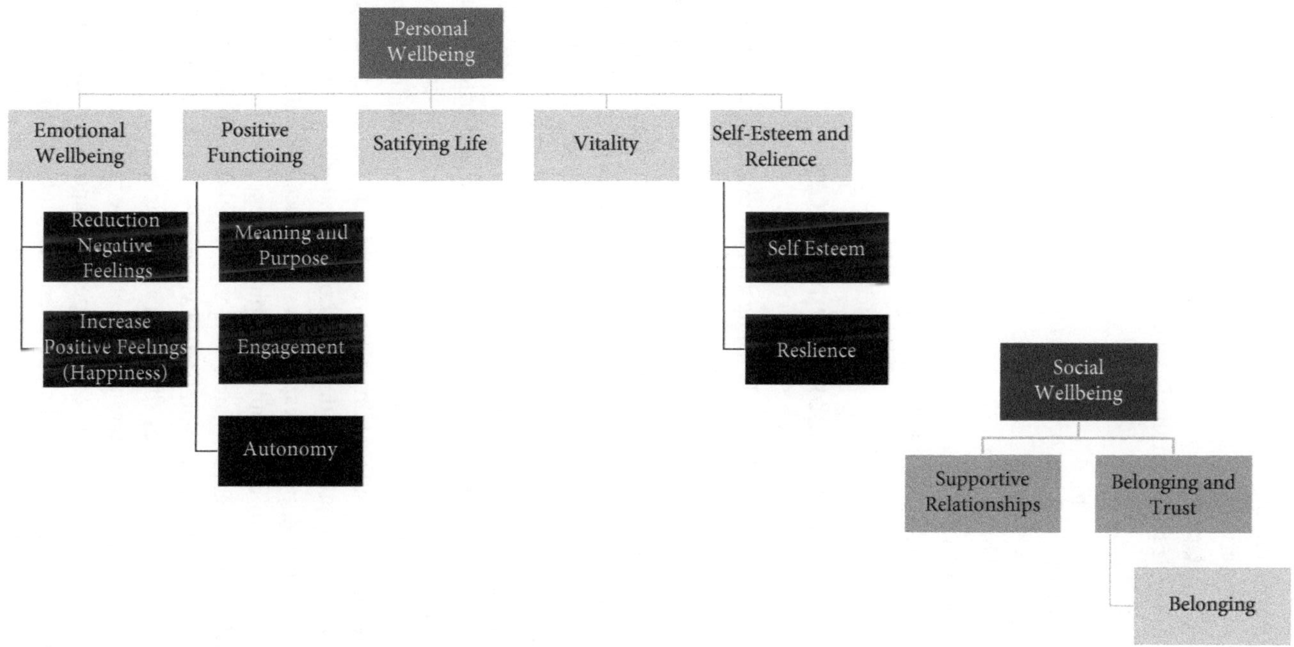

Fig. 7.1 Diagram representing how heritage places impact National Framework indicators (based on Michaelson et al. 2009).

visitors' social wellbeing, heritage places can support and promote human rights, including the right to feel connected to their own identity, to a sense of ownership, and to have an attachment to people and things (Hodder 2010). Heritage places' support for visitors' hedonic and eudaimonic wellbeing is based on the following thematically centred heritage-based pathways.

1. Interpersonal and personal experience.
2. Personal interactions.
3. Identity and belonging.
4. Rights and dignity.
5. Emotion and placemaking.

Interpersonal and Personal Experiences

The research showed that despite labelling everybody participating in this study as 'visitors', potentially homogenizing them, individuals' experience of heritage places resulted in a range of different changes to their hedonic and eudaimonic wellbeing (Prentice 2001, Prentice et al. 1998, Smith 2004). These individual experiences of heritage sites that affected visitor wellbeing were impacted by the capabilities of the visitors, the opportunities provided by the site, visitors' motivations for visiting the site, and by the physical and social-cultural context in which the site is situated. A homogenizing approach to visitors by professionals at heritage places, including a standardized approach to heritage interpretation, created conscious and subconscious barriers in visitor engagement, including visitors' sense of belonging to and feeling of ownership of heritage places and peoples. This narrow understanding of who the 'visitor' is results in some heritage sites creating passive experiences, uncritically creating singular narratives that seek to create authentic physical experiences of heritage places in which interpretations are based on a singular dominant national narrative, which had a negative impact on visitors' wellbeing (Smith 2006). For example, specific case study sites, such as historic estates, often focused on white-centric narratives that inadvertently catered for traditional audiences. This had negative impacts on the personal and social wellbeing of some individual visitors and has resulted in some demographics, for example under 30-year-old visitors to US heritage sites, feeling excluded, and as a result their visits to these heritage places negatively impacted their wellbeing.

This research demonstrated that visitors were critically engaged in and with heritage places and the narratives they present. Individuals taking part in this study expected and wanted, through their encounters with heritage places, physical and psychological experiences that are authentic 'to them' and their perceptions of history (Prentice 2001, Prentice et al. 1998, Smith 2004). Research demonstrated that when heritage places support emotional connections with physical places and environments, and support connections between people in the present and those in the past, they positively impact visitors' sense of identity, create new memories, and improve a sense of belonging. By supporting these attributes, heritage places can improve visitors' hedonic and eudaimonic wellbeing.

Personal Interactions

Visitors' interactions and experiences with other people at heritage places, including other visitors, staff, and volunteers, had a significant impact on their personal and social wellbeing. Visitors who experienced positive interactions with staff and volunteers and who were engaged in conversations felt they had shared experiences, were made to feel welcome, and felt that they had received hospitality. These positive personal and social interactions improved visitors' wellbeing, including their social wellbeing, building a sense of belonging and trust, and developing new supportive relationships.

Visitors' personal interactions and perceptions of these personal interactions at heritage places were influenced by Covid-19 and post-Covid restrictions. Restrictions in some heritage places, including restricted access to areas of sites which reduced freedom of movement around places and the removal of interpretative material, impacted visitor autonomy and negatively impacted their wellbeing. This was particularly evident at Chatsworth and Biltmore Estate. Visitors reported that their perceptions after Covid-19 of crowding, high visitor numbers, and limited autonomy were negative experiences that resulted in them feeling more afraid and upset. These jarring experiences at heritage places, where visitor expectations, perhaps of freedom and space, did not correspond to actual visitor experiences, had a negative impact on visitors' emotional wellbeing and hedonic wellbeing. This study indicated that negative social interactions linked to social wellbeing also influenced visitors' personal emotional wellbeing, including feelings of happiness, suggesting the two are entwined.

The research highlighted that visitors' personal interactions at heritage places often failed to support belonging and trust. This suggested that for heritage places to support visitors' belonging and trust to a place and people, and to develop supportive relationships between visitors, communities, and staff, they should consider providing spaces and places indoors and outdoors for visitors, staff, and volunteers to have meaningful, relaxed, friendly, and conversational interactions with each other. This requires staff and volunteers to work together to develop visitor interactions which are personal, conversational, and collaborative.

Meaning, Identity, and Belonging

Visiting heritage places is part of a wider cultural process and supported individual visitors to make meaning about the present through the past. These processes enabled visitors to create and reinforce personal, family, and national identity, and improve personal wellbeing, in the component of positive functioning and social wellbeing, and in the component of trust and belonging. Research indicated that despite academic and professional discourse highlighting the benefit of heritage places interpretating the past through narratives and storied experiences that create personal and emotional connections, heritage interpretation at these case study sites often did not consider the potential emotional experiences and connections of material presented to visitors (Uzzell 1989). The majority of heritage places in the case studies focused on factual evidence as perceived by professional, objective, physical experiences. This hindered heritage places from improving visitor personal and social wellbeing, including individuals' self-esteem and resilience (Uzzell 1989).

Heritage places, including historic estates and historic buildings, often focused on traditional 'owner'-based white-centric narratives that excluded dialogues associated with personal memories, personal stories, and empathic human stories. It was the exclusion of emotive stories and broader interpretations that limited heritage places impact on visitors' social wellbeing, and their sense of belonging and trust to each other and to the wider community. The exclusion of or limited narratives about historic trauma or historic privilege, including slavery, domestic servitude, domestic violence, race, and gender, potentially resulted in visitor distrust and dislike by some visitors regarding the narrative of ostentatious display of wealth at some historic estates. This resulted in visitors expressing guilt, which was often linked to perceptions of historic privilege, and inhibited visitors' ability to develop personal connections and a sense of belonging to heritage places that would positively improve their wellbeing.

Research indicated that heritage interpretative strategies create distance and arbitrary divisions between visitors and heritage professionals, and this hinders visitors' emotional connections to people and places, which can negatively affect personal and social wellbeing (Uzzell 1989). This included visitors feeling lack of choice and autonomy, freedom of movement around a site, feeling upset or hostile, and feeling their visit was not worthwhile. It seems likely that by not presenting diverse stories that are connected to personal, family, or national identity this had measurable negative impacts on visitors' hedonic and eudaimonic wellbeing, including their emotional wellbeing and self-esteem and resilience, as visiting some heritage places increased feelings of upset, guilt, being afraid and hostility, and feeling life not worthwhile.

Human Rights and Human Dignity: Equality, Diversity, and Accessibility

The socio-economic situation of the country in which heritage sites are located, together with their particular location within the country, had a measurable impact on heritage sites' ability to positively change visitors' wellbeing. Indian heritage sites located in areas of high poverty and in poor communities had the most measurable and significant impact on visitor wellbeing. This was linked to providing outdoor space and informal learning spaces to improve personal wellbeing, including emotional wellbeing and positive functioning. This suggests that heritage sites have a role to play in reducing wellbeing inequality both locally, nationally, and internationally.

Covid-19 increased wellbeing inequality and had a disproportionate impact on young and older populations, disabled people, Black, Asian, and minority ethnic individuals, and poorer communities (https://whatworkswellbeing.org/resources/the-impact-of-the-pandemic-on-subjective-wellbeing-inequalities/). The impacts of Covid-19 included a reduction in socialization, communication, emotional and mental health, low levels of physical activity, and increased sedentary behaviour (https://whatworkswellbeing.org/resources/the-impact-of-the-pandemic-on-subjective-wellbeing-inequalities/). Results from this research indicated that the age, gender, and disability status of visitors impacted how heritage affected their personal and social wellbeing. Heritage places significantly improved the social wellbeing of visitors with disabilities, and helped them to connect, an improvement not identified in able-bodied visitors. Heritage places had positive impacts on supporting individuals in the United Kingdom who were over 60 to learn, and individuals under 30 to be active. Heritage places also improved the emotional wellbeing of visitors in the United States in older age groups, (40–49, 50–59, and over 60), specifically reducing feelings of distress after visiting. Results indicated that females aged 40–49 with no disability experienced the greatest positive change to their personal and social wellbeing. Heritage places did not improve the social wellbeing of visitors aged 30–39 and 50–59 (UK), and males and those under 30 (US), and had a negative effect on the personal wellbeing of women and visitors under 30 at US sites. The research has highlighted that the impact on visitors of heritage sites varies with visitor demographics. It also demonstrates the potential of heritage places to reduce the wellbeing inequality of disabled, young, and elderly populations, and poorer communities.

Heritage site visitor demographics and study participant demographics indicated low levels of ethnic diversity and disability amongst visitors. Many heritage sites did not, as previously noted, address contested heritage, and there was an absence of dialogues about historic trauma or historic privilege. By not presenting inclusive and diverse narratives, by not supporting access to places, areas, or material, and by not enabling freedom of movement and freedom of choice for all potential visitors, heritage places limit their ability to increase visitor numbers and to motivate a wider demographic of visitors to engage with them. This in turn reduces their ability to have a positive impact on wider wellbeing equality and to improve wider societal wellbeing.

The research demonstrated how respect for human rights is connected to wellbeing indicators, including the New Economics Foundation five ways to wellbeing and the National Accounts for Wellbeing (Aked et al. 2008, Michaelson et al. 2009). These connections between human rights and wellbeing are linked to individuals' personal functioning and functioning behaviour, for example the right to feel connected to their own identity, to feel a sense of ownership and belonging, to have an attachment to things and people (to connect), the right to education (to keep learning), and the right to think, to feel, to imagine (to take notice) (Hodder 2010: 872, Aked et al. 2008). This study indicated that access to and participation in heritage positively impacts wellbeing, and access to heritage should be considered, as Hodder suggested, a human right (Hodder 2010). Therefore, it could be considered that reducing wellbeing inequality and improving visitor and wider community wellbeing should be an ethical and moral duty of heritage professionals that work at, for, and with heritage places.

Emotions and Placemaking: Therapeutic Landscapes

The power of cultural and heritage landscapes to improve subjective wellbeing, including that of people with pre-existing mental health problems, has been evidenced by numerous studies (Darvill 2019, Clifton 2019). These studies suggest cultural landscapes provide therapeutic experiences and transitional spaces to support physical and mental wellbeing and spiritual healing. This has been linked to heritage landscapes supporting emotional connections, enabling people to develop their own cultural meanings, their sense of belonging, and personal relationships to places and people (Darvill 2019, Clifton 2019, Waterson 1994: 45, Tilley 1994).

Research demonstrated the physical aesthetics of the heritage place, including its presentation and conservation, impacted visitors' wellbeing, most notability visitor personal wellbeing. For example, the poor preservation and 'disregard' or lack of perceived protection for physical remains at some heritage place negatively impacted visitor wellbeing. Heritage places that provided outdoor environments for visitors to explore and experience supported hedonic wellbeing, specifically improvement in positive feeling linked to 'happiness'. This research enabled a deeper understanding of the specific attributes of wellbeing that outdoor spaces supported. For example, heritage places with gardens and landscapes provided visitors with access to external 'outdoor' physical spaces and wider cultural environments to explore, such as historic gardens, lakes, woods, and landscapes. These aspects measurably improved visitors' personal and social wellbeing. Wider external heritage landscapes supported emotional wellbeing and happiness and had a 'therapeutic' impact on visitors' mental health and wellbeing, supporting them to feel calm, peaceful, happy, and relaxed; they provided space to freely explore and to feel autonomy and support positive functioning, to be outside in nature and support vitality, to connect them to the place, and to support a sense of belonging. This is consistent with other research ideas that heritage places and landscape can produce therapeutic benefits (Darvill 2019, Tilley 1994, Gesler 2005, Williams 2007).

The therapeutic benefits to visitors' emotional wellbeing and feelings of happiness resulting from visits to heritage places that had no or limited outdoor space, for example galleries and museums, were less significant, but still evident. Galleries had most measurable impact on visitors' personal wellbeing in the areas of positive functioning, supporting visitors to keep learning, and to have a meaningful experience. This research indicates that art and nature have therapeutic benefits, and that heritage places with green spaces and art improved visitors' wellbeing and benefited their wider mental health. These findings have wider implications for national and regional health care agendas, including the value of visiting and engaging in heritage places as a form of preventative medicine and social prescribing.

Holistic Management Models, Wellbeing Strategy, and Policy

The study indicated that the management of heritage places impacted visitors' wellbeing. Heritage places where the management was based on humanistic 'people-centred' principles and that were managed on a not-for-profit basis appeared to have greater positive impact on visitors' subjective wellbeing (Pirson 2017). For example, Indian heritage places, managed by government agencies, and not based on market models of profitability, had the most measurable and significant impact on NAF wellbeing indicators (Michaelson et al. 2009). On the other hand, heritage places that were managed on market-based models and that as a result charge higher ticket prices, for example Biltmore Estate, had lower levels of impact on wellbeing indicators and some negative impact on specific indicators, including to personal wellbeing components of positive functioning, self-esteem and resilience, and emotional wellbeing. Lower levels of measurable change in visitor wellbeing after visiting sites with high ticket prices, for example Chatsworth and Biltmore Estate, potentially related to higher visitor expectations not being met by visitor experience. The commercialization of heritage places creates a model of power based on producers and consumers of heritage and could potentially have a negative impact on visitor wellbeing, placing into question the sustainability of market-based models, if the requirement to make profits outweighs the goal to provide a positive experience to visitors.

A measure of success at market-driven heritage places is economic impact, and sustainability models and visitor strategies are often based on this (Friedman 1962). However, as demonstrated in this book the success of heritage places cannot solely be determined by economic models, as heritage places that applied holistic or people-centred approach (such as DPG and CGMAG) were more successful in improving visitors' wellbeing. Wellbeing is part of human and social capital but has also been linked to economic capital; a management approach that places wellbeing at the centre of strategies could have a positive impact on economic, human, and social capital and support the sustainability of heritage places and communities (Pirson 2017, Scott 2012, Hodder 2010, Lassibille and Gómez 2014). Measuring the success of heritage places based on the benefits to social and human capital as well as to economic capital would provide evidence of heritage places' role in sustainable development, levelling up, tourism, health, equality, diversity, and inclusion (EDI), and human rights agendas, to name but a few (UK Government Housing and Communities 2022, UN 2015, UNESCO 2016, Council of Europe 1950, OECD 2001, Sagger et al. 2021, WHO 2022).

This research has indicated that heritage places can both support local and national wellbeing and health agendas, and can also support global agendas, such as the UN's 2030 sustainable development goals, including the ability to positively impact SDG3: Good Health and Wellbeing; and in some circumstances support SDG4: Quality Education; and they might have the potential to support SDG5: Gender Equality (United Nations 2015, Harrison et al. 2020, UNESCO 2016). Evidence presented in the book suggests that heritage places could and should do more to support good health and wellbeing, wellbeing inequality, and wider human right goals, including those linked to SDGs (Council of Europe 1950, United Nations 2015, Abdallah et al. 2017).

This book has presented potential strategies by which heritage places could support these human rights goals by taking a people-centric management approach that places wellbeing at the heart of strategy, and supports the development of complex and meaningful relationships, interactions, and interpretation with visitors, the public, and the community to support human dignity, identity, belonging, connections, education, enjoyment, and meaning (Pirson 2017). This approach might provide capabilities, opportunities, and motivations to enable, encourage, and facilitate wider community access to heritage places, which is arguably a human right, and facilitate diverse communities to engage and feel a sense of belonging, connection to, and ownership of heritage place (United Nations 2011, Giberito 2021).

To improve visitor and community wellbeing, heritage places might consider a holistic approach to management that combines humanistic and economic models, supports wellbeing, and provides (Pirson 2017):

1. The economic capital to conserve and preserve heritage, support visitors' aesthetic physical experiences, and provide facilities and access to support the physical capabilities of people to visit, supporting people to 'take notice and to be active' (Sagger et al. 2021).
2. The social and human capital to create meaningful 'people-centred' storied experiences, to provide autonomy, to increase the motivation for people to visit, supporting people 'to connect', 'to give', and 'to keep learning' (OECD 2001, Sagger et al. 2021).

Future Strategies to Societal Wellbeing Value Chain

Heritage places positively impacted the wellbeing and mental health of visitors. Results suggest that all heritage assets needed to develop visitor experiences that support visitors to keep learning, to connecting, and to giving. Heritage places provide potential mechanisms for health care professionals and government agencies to strengthen their wellbeing and mental health provisions for the benefit of individuals and communities. It is important for the future sustainability of heritage places that future strategies that support wellbeing are implemented. These include:

- Develop provisions to support positive functioning.
  - Develop new engagement activities, including learning/interpretative resources, audio guides, guided tours, and storied experiences for visitors, to enable them to keep learning about their pasts and have meaningful experiences at heritage places.
  - Develop interpretative methods to support and empower visitor access to material and autonomy. Consider visitor engagement and interpretation strategies that present and empower freedom of choice in storied experience and movement around spaces and places.
  - Provide specific co-created engagement opportunities that have meaning and purpose for different demographics, that provide opportunities to empower specific communities to keep learning

about their history and heritage, and that provide autonomy that creates ownership and drives projects and activities forward and has a legacy.

- Support emotional wellbeing, specifically personal 'happiness'.
  - Develop programmes which utilize arts and creative cultural activities to improve positive feelings, confidence (self-esteem and resilience), and emotional awareness through self-reflection.
  - Invest in outdoor spaces and landscape that develops outdoor areas for different connections and reflections and that encourages personal reflection, happiness, and therapeutic benefits of natural landscapes and nature. Use outdoor spaces to support changing attitudes, values, and behaviours of visitors in how they use and act in outdoor space to improve wellbeing.
  - Develop comfortable indoor spaces, with seating and with views or creative material for families to connect, talk, relax, and reflect.
  - Partner with mental health charities and organizations within the local community, to take part in mental health awareness programmes and support community access to heritage places and tools to improve mental health and wellbeing. This might involve working through social prescribing programmes and NHS link-workers in primary care networks.
  - Preserve, conserve, and take care of physical heritage, to support positive authentic aesthetic experiences for the visitor and reduce negative impacts to wellbeing.
  - Provide staff and volunteer training to support communication skills and visitor hospitality. Engender a friendly, kind and open approach to all visitors to reduce negative feeling and emotions.
- Support vitality and physical wellbeing of visitors.
  - Provide specific activities, time, and space for visitors to be outdoors and be active.
- Support self-esteem and reliance and life satisfaction.
  - Support activities on site for visitors to undertake ecological awareness activities, to improve life satisfaction.
  - Develop cultural activities linked to wider social, environmental, and economic inequality, to encourage visitors to give something back to the community (such as giving up time, or expertise), to enable people to feel worthwhile, build self-esteem and resilience, pride, and improve life satisfaction.

Continue to build strengths in supporting social wellbeing.

- Continue and strengthen heritage places' relationships with the community and community organizations to support social wellbeing, and to develop trust and a wider sense of community belonging with and in heritage places. Enable heritage places to play an active role in regional agendas, including localization agendas and to support local community cohesion and identity.
- Develop events to connect visitors, staff, volunteers, and community members with each other to help solidify supportive relationships and support local community cohesion.
- Create areas and spaces that encourage conversations and interactions between employees/volunteers/staff/visitors/community members to enable them to connect and share experiences and knowledge with people around them.
- Co-create storied experiences with visitors and communities to develop trust and a wider sense of community belonging with and in histories presented at heritage sites. This might also support wider localization agendas and local community cohesion.
- Adapt dialogues and narratives to reposition interpretations, from sanitized narratives perceived as pleasant by visitors, to presenting community-driven stories of the hidden histories of the heritage sites that link to personal, family and community, and national identity, and that improve belonging.
- Promote and market the holistic and wellbeing benefits of visiting heritage places to break down preconceived barriers to visiting buildings and spaces.
- Continue to develop in-person led tours to support people to connect with each other.
- Use external and internal spaces to creatively support learning and as active spaces to build relationships and belonging and trust.
- Provide a time and space for diverse groups and individuals to co-develop new interpretations and storied experiences. Employ creative arts methods (use of photography, audio, moving image, and mapmaking) to present community-led interpretations through performance-based narration of storied experiences.
- Consider free community passes or transportation for local residents to encourage more diverse visitors and reduce wellbeing inequality.

To support wellbeing requires heritage places to support visitor experiences that improve identity, and support connections, belonging, and emotional connections (Figs 7.1 and 7.2).

Value Chain Model and Visitor Experience Model

The idea of a value chain and visitor experience model is to provide a framework that supports heritage places to design visitor experiences aimed at improving visitors hedonic and eudaimonic wellbeing outcomes, yet more broadly relates to developing a societal value chain model for heritage places. The societal value chain model for heritage places enables professionals understand the current social impacts of visiting and based on this knowledge to design new strategies for visitor engagement based on the ideal visitor experience that seeks to impact visitors' feelings and capabilities to improve individual wellbeing (Fig. 7.2). Based on this societal value chain model (Strategies & Visitor Engagement Design > Ideal Visitor Experience > Visitor Engagement Impact (to feelings) > Individual Wellbeing > Societal Wellbeing) heritage places can develop a business model based on how, by positively impacting visitors and community wellbeing, people will be more likely to return to heritage places and

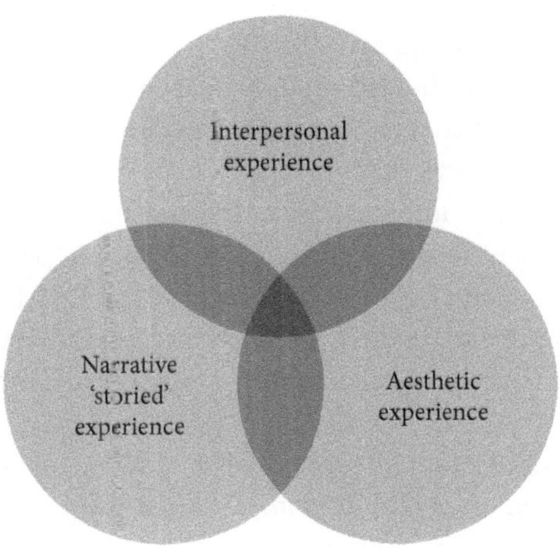

Fig. 7.2 How to create visitor wellbeing: visitor experience model for wellbeing.

visitor numbers will increase as the site appeals to broader demographics. Not only would this approach improve individual wellbeing, but it would also support societal wellbeing and reduce wellbeing inequality. This societal value chain model enables heritage places to improve their economic capital by supporting social and human capital (Sagger et al. 2021).

Behaviour Change Model

The themes identified in this research and framework for improving individual, community and social wellbeing provide the possibility for behaviour change models to be applied to wellbeing and heritage place research to shape strategy recommendations and heritage place interventions, and to enhance the ability to support policy making and the integration of heritage places into policy (Gallou 2022, Michie et al. 2014a, 2014b, Michie et al. 2011). For example, the COM-B behaviour change model has been used in a range of health policy interventions and provides a potential mechanism to plan and implement strategies based around key components identified in this research that are critical to heritage places improving wider societal wellbeing (Michie et al. 2014a, Michie et al. 2011). These are:

- Capability, split into the subcategories of physical capability and psychological capability and potentially covering themes such as equality, diversity, and accessibility.
- Opportunity, sub-divided into physical and social, which here helps to frame concepts such as personal interactions, and interpersonal and personal experiences.
- Motivation, split into automatic and reflective and relevant to the themes of identity and belonging, emotions, and placemaking (Cane et al. 2012, Gallou 2022). According to the COM-B model, these three components outline the necessary conditions required for a behaviour to occur (Michie et al. 2011).

The COM-B model offers the potential to achieve high-level outcomes through behaviour change, for example 'more people come to heritage places' and 'a greater diversity of visitors engage in heritage places'. The outcome of more people coming to heritage places might be that the engagement of higher levels of the population in heritage places results in a higher number of individuals' wellbeing being improved, and heritage places are able to positively impact societal wellbeing. The advantage of the COM-B model is in both its simplicity and specificity, i.e. it can be adopted by individual heritage places to target specific audiences at specific sites. This model can therefore be used to design strategies to encourage more diverse participation in heritage, and hence to use heritage places to target the reduction of wellbeing inequality. However, specific interventions for particular heritage places and audience's behaviours will require a clear understanding of who (needs to perform the behaviour), what (that the audience needs to do differently to achieve the desired change), when (they need to do it), where (they need to do it), how often they should do it and with whom (they should do it) (Michie et al. 2014b). COM-B model can be used as a framework to better understand the barriers that prevent people from visiting heritage places (Fig. 7.3).

The COM-B model forms the central part of the Behaviour Change Wheel (Michie et al. 2014b). This expanded model helps to identify different intervention types to effect different barriers/enablers to a given behaviour. For example, 'environmental restructuring' is an effective intervention to increase both physical opportunity, social opportunity and automatic motivation towards a given behaviour. Using COM-B model enabled the creation of a pathways to heritage site wellbeing framework, through the lens of the behaviour change wheel intervention types, which built on the information presented in Figure 7.4. The following four intervention types are relevant to heritage site wellbeing: education, persuasion, enablement, and environmental restructuring (Fig. 7.4).

Although not used here to design a specific intervention, and therefore not developed to the stage of mapping behaviour change techniques to these interventions, the COM-B/BCW model has been briefly referred to in this book to highlight potential heritage wellbeing applications (Michie et al. 2014a, 2014b, Michie et al. 2011). Further research and the identification of specific localized and contextualized case studies is required to test the potential of this model for use by heritage professionals. Figure 7.4 provides a demonstration that COM-B/BCW can offer an alternative way of framing the theories of wellbeing and heritage places identified in this book, which highlights the potential value of the science of behaviour change to the heritage sector.

High level objective	Desired behaviour change(s)	COM-B	Description
Heritage sites impact societal wellbeing	Increase visitor numbers and diversity of visitors to heritage sites	Capability	Physical People lack physical strength and ability to visit heritage places. Psychological People do not know the social benefits of visiting heritage places. People do not feel supported, emotionally, to visit heritage places. People do not think they have the skills to learn or knowledge to understand information presented at heritage sites. People do not know how to get to heritage places and were to find visitor information.
		Opportunity	Social Heritage places are only visited by a narrow demographic of society. Physical Lack of physical access to areas of sites or sites themselves. Lack of time to visit heritage place. Lack of money to visit heritage sites. Insufficient parking at heritage sites or transport to heritage sites. Lack of facilities for visitors at heritage sites.
		Motivation	Automatic Heritage sites are perceived as boring and associated only with 'formal' learning. Visiting heritage places is not part of leisure and self-care habits of all communities. Reflective Visiting heritage places are not part of people's goals to improve wellbeing or connect to people and places. Heritage places are not perceived as representing self-identity. Heritage places are not perceived as relevant to communities. Communities do not feel they will learn about their heritage, culture, or traditions.

Fig. 7.3 Table demonstrating how COM-B model can be used as a framework to better understand the barriers that prevent people from visiting heritage places.

Pathways to Heritage Site Wellbeing

Intervention Type	Intervention Strategies	Thematic Framework
Education	Educate wider community about the benefit of heritage sites through marketing and online material.	Psychological capability: human rights, human dignity, equality, diversity
	Schools and community educational programmes based around heritage sites health and wellbeing benefits.	Physical capability: accessibility
	Promotional material for healthcare community to support referral patients to heritage sites as part of health provisions.	Social opportunity: personal interactions, interpersonal and personal experiences
	Create on-site health and wellbeing learning programmes and interpretative material.	Reflection motivation: identity and belonging and placemaking
	Use creative arts-based community led interpretations through performance-based narration of storied experiences (photography, audio, moving image, and mapmaking).	Automatic motivation: emotions
	Provide ecological and environmental awareness activities.	
Modelling	Community access programmes to change perception of heritage places to diverse communities.	Reflective motivation: identity and belonging, and placemaking
	Appoint 'community advocates' to highlight social benefits of heritage places.	Automatic motivation: emotions
Training	Provide staff and volunteer training to support communication skills and visitor hospitality, to encourage friendly, kind and open approach to all visitors.	Psychological capability: human rights, human dignity, equality, diversity Physical opportunity: accessibility
	Partner with mental health charities and organisations within the local community, to take part in mental health awareness training programmes and support community access to heritage places.	Automatic motivation: emotions

Fig. 7.4 Table presenting COM-B interventions as pathways to heritage site wellbeing. Table presenting pathways to heritage site wellbeing framework, through the lens of the behaviour change wheel intervention types.

Pathways to Heritage Site Wellbeing

Intervention Type	Intervention Strategies	Thematic Framework
Persuasion	New marketing and media adverts to realign heritage places with social values and as community assets.	Reflective motivation: identity and belonging, and placemaking Automatic motivation: emotions
Enablement	Free community passes and/or free transportation for local residents Use external and internal spaces to be active learning spaces. Develop in-person led inter-generational tours. Adapt dialogues and narratives to present community driven stories of the hidden histories and different identities. Co-create opportunities to empower communities to present their history and heritage. Host socially based community events for visitors, staff, volunteers, and community members to come together and share experiences. Develop cultural activities linked to wider social, environmental, and economic inequality. Provide specific activities, time, and space for visitors to be outdoors, move, and be active. Develop programmes which utilize arts and creative cultural activities. Create new engagement activities such as interpretative resources, audio guides, guided tours, and storied experiences for visitors.	Psychological capability: human rights, human dignity, equality, diversity Physical opportunity: accessibility Social opportunity: personal interactions, interpersonal and personal experiences Automatic motivation: emotions

Environmental Restructuring	Through public transport or free electric transport enable ease of access to and freedom of movement around heritage places. Provide (if possible) disabled access to enable movement around entire site. Provide material in different formats that support all individuals' diverse learning and behavioural needs. Develop comfortable indoor spaces, with seating and with views or creative material for people to talk, relax, and reflect. Place wellbeing at heart of organizational, learning, interpretation, and visitor experience strategies. Develop relationships with the community and community organizations through collaborative partnerships, free access, and playing an active role in regional agendas, including localization agendas and community cohesion agendas. Preserve, conserve, and take care of the physical heritage. Invest in and integrate outdoor spaces and landscapes within tours and active interpretations. Develop outdoor provisions, areas for different connections and reflections. Develop interpretative strategies and wayfaring that empowers visitor freedom of choice in movement around site and access to cultural heritage material. Develop volunteers' network to support people to provide their time and expertise to support community assets. Develop spaces that encourage conversations and interactions between employees/volunteers/staff/visitors/community members.	Physical opportunity: accessibility Social opportunity: personal interactions, interpersonal and personal experiences Automatic motivation: emotions

Fig. 7.4 *Continued*

Evaluation at the Heart of Practice

Since the research on which this chapter is based was started, there have been developments in wellbeing measurement and evaluation (Fancourt et al. 2020, Fancourt and Finn 2019, Bonacchi et al. 2021). This has included more varied approaches to evaluating the impact of archaeological projects, heritage sites, projects, and museums on visitor and participant wellbeing, but there is still a lack of agreement about how to collect, analyse, and present the data (Brizi et al. 2023, Sofear et al. 2021, Everill et al. 2020, Everill and Burnell 2022, Price and Keynes 2020, Gallou 2022, Cutler 2022, Heritage Alliance 2020, Historic England 2022, Damaso et al. 2023a, 2023b). What has changed however is that heritage places and projects now understand the need to place evaluation at the heart of practice, rather than as an afterthought. To make this a reality, evaluation frameworks for measuring wellbeing change must be developed in tandem with project planning, strategy, and design.

Recently developed wellbeing measurement tools, such as the Wellbeing Numerical Rating Scales (WB-NRSs)—which was designed for use to measure change in wellbeing in non-health care settings—have scope for use within heritage-based locations (Bonacchi et al. 2021). These new measurement tools, which have been developed by psychologists, and some of which have been validated in multiple languages, offer the potential for use in the collection and analysis of comparative international and cross-sector data on the impact of heritage places on wellbeing (Bonacchi et al. 2021). To assess the value of these tools the first step would be to carry out a pilot study within a community and heritage setting, considering the disciplinary, contextual, and demographic validity of the tools using a large sample size.

The present research has gaps in data collection, including the geolocational data of visitors. There is also a need for more varied demographic data to be collected to understand the value of heritage places to and within local communities. This could be achieved through the mapping of the distance of visitors' places of residence from the heritage site, with the aim of assessing whether the distance of a visitor's residence from the site (local, national, or international location) has an impact on specific wellbeing components. This would make it possible to analyse the differential impact of local sites to a community's feeling of connection, belonging, and identity.

Critical to the provision of further robust evidence of how heritage places can improve local visitors' wider community wellbeing is the collection of baseline data from non-heritage visitors and heritage visitors from within the wider local community. Future studies require the collection of benchmark data from a control group within the community, together with data from a community-based experimental 'visitor' group. This will make it possible to establish causality of results by being able to isolate the independent variable (i.e. visiting a heritage site).

The methods outlined in this book for heritage evaluation approaches should not be set in stone; rather they should be used to provide a framework for discussion to encourage critical self-reflective adaption and an input to the continuous development of evaluation methodologies. For example, one area to consider would be development of additional questions to measure 'to give'. Another idea was to develop a shorter MWS 'postcard' survey for children, and to trial alternative methods within heritage settings linked to measuring change not only in wellbeing but also emotional intelligence and literacy, which might include Stirling Children's Wellbeing Scale or the Emotional Literacy Scale (Liddle and Carter 2015, Alemdar and Amlan 2020). It would also make sense for measurements to wellbeing to be adapted for each location's particular characteristics and based on people's lived experiences of wellbeing (Scott 2012).

What the research has demonstrated is the requirement for a tailored approach to the evaluation of the impact of heritage on individuals' wellbeing, grounded in collecting robust empirical evidence of change. More evaluation work is needed to capture the long-term impact of heritage places on visitors and their legacy to communities, including whether changes to wellbeing are transient (short term), medium term, or long term. Despite the complexities of collecting this data from large data sets and the level of detail required to provide evidence of change in wellbeing based on specific interventions, for example visiting a heritage site, robust evaluation is critical if heritage places are to be integrated into wider future strategic plans and policies, including education, health, and tourism.

Recommendations for Future Research

The research suggested links between heritage, wellbeing, and human dignity, yet further exploration is required as to the role of heritage in supporting human dignity and more broadly human rights (Hodder 2010).

This includes research into the nature of the relationship between wellbeing, human dignity, and human rights and the role that heritage places can play through wellbeing provision in supporting wider equality, diversity, and inclusivity agendas.

This study indicated the potential positive impact heritage places on human dignity, specifically related to ethical practice, human rights, identity, equality, and inclusivity. However, there is no standardized measure to assess the impact heritage sites have on human dignity, beyond those associated with general wellbeing. This means at present research is unable to provide evidence of the complex and nuanced emotional reactions in relation to the feelings of dignity that visitors experienced. A potential way forward might be through incorporating key elements/principles of the Human Dignity Index (worth, recognition, and response, autonomy, safety, and security), to identify the nuanced impact of heritage sites on human dignity and the distinct elements of belonging and pride (Human Dignity and Humiliation Studies 2008). Understanding human dignity might support engagement and interpretation strategies, including storied experiences, and help professionals integrate contested, diverse, and traumatic storied experiences in order potentially to neutralize or have a positive impact on aspects of human dignity and wellbeing associated with historic trauma.

Future research should consider how improving and supporting individual and wider community wellbeing can be repositioned from presenting sanitized 'storied experiences' perceived as pleasant to the visitors of heritage places, to presenting interpretative stories that consider the contested heritage of historic places; for example, the potential traumatic and upsetting aspects of heritage (Brave Heart and DeBruyn 1998). The presentation of negative and traumatic stories at heritage sites might not negatively affect wellbeing but could be a mechanism to improve visitors' wellbeing.

Critically, wellbeing research needs to include people who have not visited heritage sites, as well as community groups and diverse communities, to consider how heritage places impact individuals' wellbeing in different ways and why. This could provide unique insights to mitigate the apparent disconnect of specific groups with heritage places, to potentially bridge the participation gap, and to understand how the heritage sector can positively effect change in their communities and wider societal wellbeing (Harrison et al. 2020, Historic England 2022).

This research has focused on how the physical experiences of heritage places have affected the wellbeing of participants to specific programmes or visitors to heritage places. This focus on in-person, physical engagement with heritage has to date overlooked the impact of digital and virtual heritage engagement on wellbeing. As digital heritage is a growth area, and heritage places are investing in digital technology to reach global communities, it is important to investigate the impact of digital engagement in heritage on user wellbeing and to compare the different wellbeing impacts of physical and virtual engagement.

Conclusion

Participation in heritage activities and visiting heritage places can improve visitors' subjective wellbeing, but this is not automatic; it requires the right place management and visitor engagement strategies.

The management, presentation, and interpretation of the past at heritage places can be a source of social good and can improve visitors' hedonic and eudaimonic wellbeing, providing a 'cathartic' and therapeutic experience that is engaging, enjoyable, interesting, imaginative, entertaining, shocking, and moving (Uzzell 1989, Darvill 2019). Heritage places could do more to improve visitors' wellbeing in areas of 'belong' and 'trust', 'autonomy', 'meaning and purpose', and 'vitality' to support visitors to keep learning, be active, and connect. Improving visitor wellbeing at heritage places requires socially based strategies, to provide visitors with physical and social spaces and experiences that emotionally connect them to the people and places in both the past and the present through empathic personal stories that create memories and belonging and support their own identity. By taking a socially 'people'-based approach, heritage places can provide the visitor with meaning and purpose, support them to learn about culture, heritage, and themselves, provide autonomy, and build pride and self-esteem. A potential intervention could be changing interpretations from temporally and contextually based snapshots of the past to presenting the past as part of a larger process that still operates today, and that considers localized narratives and global perspectives that not only instruct but provoke emotional responses and support personal connections (Wallace 2004, Sayer 2018, 2022). Heritage places should seek to present narratives that create emotional responses by presenting empathic and conflicting narratives, that encourage visitors to think, to learn, to question, to understand what it is to be human, to belong, and to create memories (Smith 2021). In order to support diverse visitors' wellbeing, interpretations and narratives should be presented based on shared pasts of heritage sites. This can be achieved through visitors' and communities' active participation in developing strategies and narratives that support individual identities and enable self-determination, autonomy, connections, belonging, and attachment to things and people and values (Tacon 2019, Smith et al. 2022, Yap and Yu 2016, Gee et al. 2014). This approach could support all visitors to feel respect and to personally benefit from heritage places, and as a consequence improve their wellbeing (Hodder 2010).

The book has highlighted how heritage, identity, human dignity, and wellbeing are entangled (Hodder 2010, Sen 1993, Smith et al. 2022, Donders 2020). For example, indicators for construction and recognition of visitor identity can be linked to personal wellbeing and the components of self-esteem and resilience, and social wellbeing in the components of belonging and supportive relationships. It is these components of wellbeing that are associated with human dignity (Holtorf 2018, Tacon 2019, Fujiwara et al. 2014, Dockery 2010). The management of heritage places can help visitors to explore identities and support human dignity and can be powerful tools to improve visitor wellbeing (Ander et al. 2013, Smith et al. 2022, Smith 2021). Heritage places provide people with opportunities for physical, psychological, and social connections and for attachments to occur between people and places. The presentation of the past at these places can support the construction and recognition of an individual's identity, heritage, and memory (Cubitt 2007).

Heritage places provide opportunities to support physical, social, personal, and psychological capabilities improve visitor hedonic and eudaimonic wellbeing. The research described in this book suggests that by adapting visitor strategies and interpretations, heritage places can motivate people to change attitudes, behaviour, and actions that can improve wellbeing and in turn support wider human rights and equality, including those of access, ownership, belonging, connections, and education (Uzzell 1989, Hodder 2010). Heritage places do not need to undertake specific programmes to support visitors' wellbeing, but if they want to have a wider impact on more diverse communities' wellbeing, they need to change their models of interpretation to place wellbeing at the centre of approaches that are empathic, people-centred, and localized.

Heritage places can through their management models and strategies tackle wellbeing inequality, support wider human rights, and face global challenges, but it is key to embed heritage and cultural places into wider wellbeing and government agendas and to link them to wider human, social, and economic capital (Smith et al. 2020, Hodder 2010, Zbranca et al. 2022). Heritage professionals need to consider diverging from outstanding universal value models and value models based on physical evidence, to value models based on wellbeing and linked to human rights (Hodder 2010). This can be achieved by heritage sites focusing on strategies based on improving wellbeing- and capability-building. These wellbeing- and capability-building strategies could, through improved participation, visitor access, and access to heritage to all communities and people, create narratives based on shared pasts and improve the understanding of the legitimacy of value through collaboration (Meskell 2011, LaRoche 2005). Active participation by the public in the processes of heritage management could improve visitor capabilities and provide a place to heal, and support not only visitor wellbeing but community mental health (Meskell 2011, LaRoche 2005). This book advocates that the management of cultural heritage should place wellbeing at the heart of decision making and strategies. This wellbeing value approach, advocated by Nussbaum (1993, 2005), Sen (1993), and Hodder (2010), is grounded in supporting human rights, and considers the value of heritage as part of wider cultural processes that are vital to sustainable communities.

Heritage and Wellbeing: The Impact of Heritage Places on Visitors' Wellbeing. Faye Sayer, Oxford University Press.
 DOI: 10.1093/9780191914539.003.0007

Appendix

	Name	Management of Site	Location	Comparable Participants
Heritage Site	Chatsworth Estate	Charitable Trust	UK	107
	Biltmore Estate	Private Company	US	724
	Nemours Estate	Charitable	UK	46
	Tipu Sultan Summer Palace	Government	India	96
Archaeological Site	Priory Gardens	Government	UK	37
	Bangalore Fort	Government	India	82
Museum	Imperial War Museum Duxford	Government/Charity	UK	96
	Chandigarh Government Museum and Art Gallery	Government	India	70
Art Gallery	Dulwich Picture Gallery	Charitable Trust	UK	34
	Venkatappa Art Gallery	Government	India	63

Appendix, Fig. 1 Case study heritage places, including type, location, management, and number of comparable participants.

Please tick to indicate how much you agree with these statements at the present moment:

I feel...	Not at all	A little	Moderately	Quite a bit	Extremely
Interested in the world around me					
Connected to the people around me					
Happy with my life					
Satisfied with my life					
I have the opportunity to learn new things and skills					
A sense of accomplishment					
Optimistic about my future					
I have freedom of choice					
I can overcome challenges					
Connected to the local environment					
My life is worthwhile					
I belong to a neighbourhood					
Physically active					
People support each other					
I can trust people					

Appendix, Fig. 2 Modified Wellbeing Scale (MWS).

Question	Wellbeing Indicator	Category	Subjective Wellbeing Type
Interested in the world around	Personal	Positive functioning	Eudaimonic
Connected to the people around me	Social	Supportive relationships	Eudaimonic
Happy with my life	Personal	Emotional wellbeing	Hedonic
Satisfied with my life	Personal	Life satisfaction	Hedonic
I have the opportunity to learn new things and skills	Personal	Positive functioning	Eudaimonic
A sense of accomplishment	Personal	Positive functioning	Eudaimonic
Optimistic about my future	Personal	Self-esteem	Eudaimonic
I have freedom of choice	Personal	Positive functioning	Eudaimonic
I can overcome challenges	Personal	Resilience	Eudaimonic
Connected to the local environment	Social	Belonging	Eudaimonic
My life is worthwhile	Social	Self-esteem	Eudaimonic
I belong to a neighbourhood	Social	Belonging	Eudaimonic
Physically active	Personal	Vitality	Eudaimonic
People support each other	Social	Supportive relationships	Eudaimonic
I can trust people	Social	Trust	Eudaimonic

Appendix, Fig. 3 MWS questions corresponding personal or social wellbeing indicators and categories.

Please tick to indicate how much you agree with these statements at the present moment:					
I feel…	**Not at all**	**A little**	**Moderately**	**Quite a bit**	**Extremely**
Attentive					
Distressed					
Interested					
Alert					
Jittery					
Excited					
Guilty					
Afraid					
Enthusiastic					
Irritable					
Strong					
Ashamed					
Inspired					
Scared					
Hostile					
Active					
Nervous					
Proud					
Upset					
Determined					

Appendix, Fig. 4 PANAS survey.

PANAS Words	Personal Wellbeing Categories
Attentive	Emotional wellbeing, positive functioning
Distressed	Emotional wellbeing
Interested	Emotional wellbeing, positive functioning
Alert	Emotional wellbeing
Jittery	Emotional wellbeing
Excited	Emotional wellbeing
Guilty	Emotional wellbeing
Afraid	Emotional wellbeing
Enthusiastic	Emotional wellbeing, positive functioning
Irritable	Emotional wellbeing
Strong	Emotional wellbeing, vitality, self-esteem, and resilience
Ashamed	Emotional wellbeing
Inspired	Emotional wellbeing, positive functioning
Scared	Emotional wellbeing
Hostile	Emotional wellbeing
Active	Emotional wellbeing, vitality
Nervous	Emotional wellbeing
Proud	Emotional wellbeing, self-esteem, and resilience
Upset	Emotional wellbeing
Determined	Emotional wellbeing, self-esteem, and resilience

Appendix, Fig. 5 PANAS words and associated wellbeing indicator and categories.

Exclusion Criteria	Details
Minimum age	Exclusion if participants under the age of 18.
Consent	Exclusion if participant did not provide consent or was not able to do so.
Before and after surveys	Exclusion if participants did not complete both before and after surveys.
All answers	Exclusion if participants did not complete all PANAS or MWS surveys elements.
Online surveys	Exclusion if participants did not complete after survey within the required time frame (max. seven days).
Participant numbers	If a place did not meet the required thirty-five participants completing before and after surveys, that site would be removed as it would not have been possible to undertake a statistically significant analysis.
Covid restrictions	If a place did not meet health and safety requirements or was impacted by Covid restrictions, then that place was removed.
Permissions	If project team did not get the relevant permission from project team site organizations, then that place was removed.

Appendix, Fig. 6 Exclusion criteria for case studies and data.

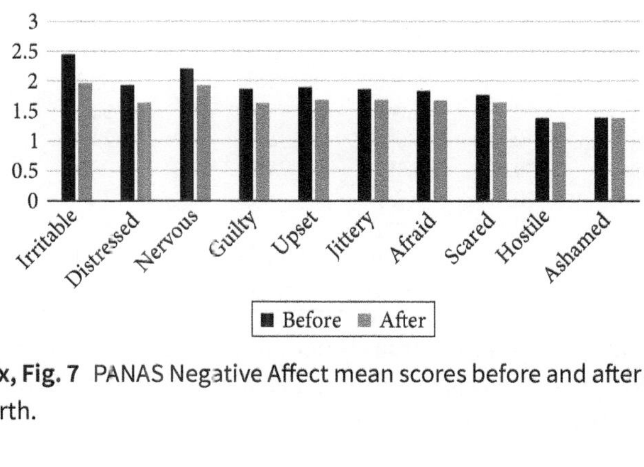

Appendix, Fig. 7 PANAS Negative Affect mean scores before and after visit to Chatsworth.

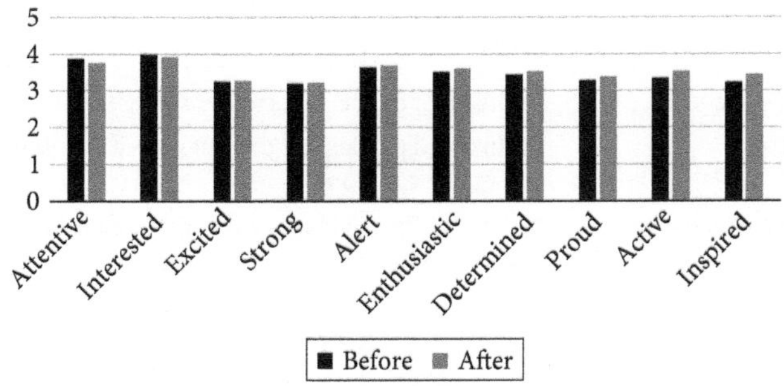

Appendix, Fig. 8 PANAS: Positive Affect Scale mean scores before and after visit to Chatsworth.

Measure	P Value	SD Change	SD % Change	Mean % Changed
Attentive	0.097	−0.012	−1.63	−2.88
Distressed	*0.000	−0.006	−0.80	−15.11
Interested	0.360	0.012	1.55	−1.51
Alert	0.775	−0.031	−3.67	1.65
Jittery	*0.003	−0.016	−1.88	−9.22
Excited	0.779	0.096	10.19	1.06
Guilty	*0.012	−0.128	−13.67	−12.04
Afraid	*0.045	0.012	1.38	−7.98
Enthusiastic	0.312	0.054	6.48	2.68
Irritable	*0.000	−0.094	−10.70	−19.65
Strong	0.583	0.074	7.96	1.08
Ashamed	1.000	−0.014	−2.02	−0.62
Inspired	0.088	0.056	5.80	6.91
Scared	*0.037	−0.043	−4.76	−6.31
Hostile	0.310	−0.033	−5.36	−2.88
Active	*0.006	−0.105	−9.29	6.19
Nervous	*0.005	−0.047	−5.12	−12.11
Proud	0.153	−0.027	−2.57	3.94
Upset	0.059	−0.038	−4.62	−10.54
Determined	0.136	−0.041	−4.06	3.00

Appendix, Fig. 9 PANAS results before and after visiting Chatsworth, including mean percentage change, standard deviation, standard deviation mean percentage change, and P value.

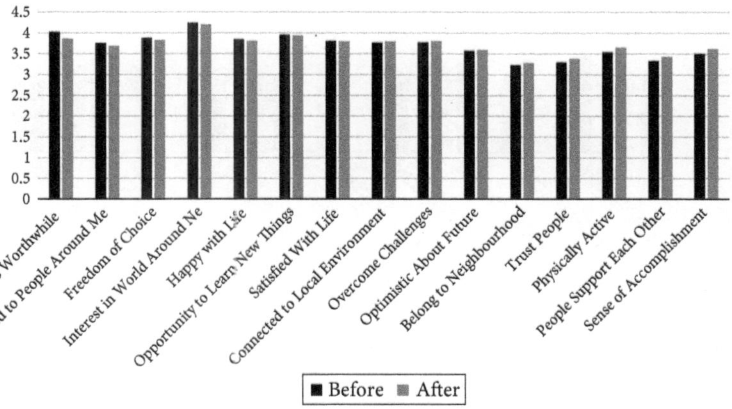

Appendix, Fig. 10 Modified Wellbeing Scale mean results before and after visiting Chatsworth.

Measure	P Value	SD Change	SD % Change	Mean % Change
I feel interested in the world around me	0.486	−0.028	−4.10	−0.81
I feel connected to the people around me	0.243	−0.041	−5.32	−1.60
I feel happy with my life	0.690	0.012	1.86	−0.67
I feel satisfied with my life	1.000	−0.061	−8.27	0.00
I feel I have the opportunity to learn new things	0.868	−0.018	−2.31	−0.22
I feel a sense of accomplishment	0.055	−0.018	−2.23	3.43
I feel optimistic about my future	0.728	−0.066	−7.58	1.20
I feel I have freedom of choice	0.755	0.032	4.02	−1.32
I feel I can overcome challenges	0.522	0.030	4.14	1.14
I feel connected to the local environment	0.855	−0.064	−7.54	0.91
I feel people support each other	0.221	−0.036	−4.30	3.35
I feel I can trust people	0.220	0.028	3.59	2.59
I feel my life is worthwhile	*0.007	−0.037	−4.22	−4.04
I feel physically active	0.417	−0.096	−9.16	3.15
I feel I belong to a neighbourhood	0.770	0.006	0.55	1.86

Appendix, Fig. 11 MWS results before and after visiting Chatsworth, including mean percentage change, standard deviation, standard deviation mean percentage change, and P value.

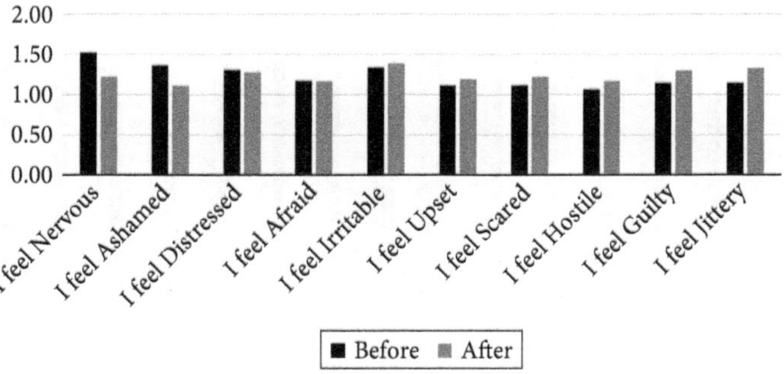

Appendix, Fig. 12 PANAS negative affects before and after visiting Priory Gardens.

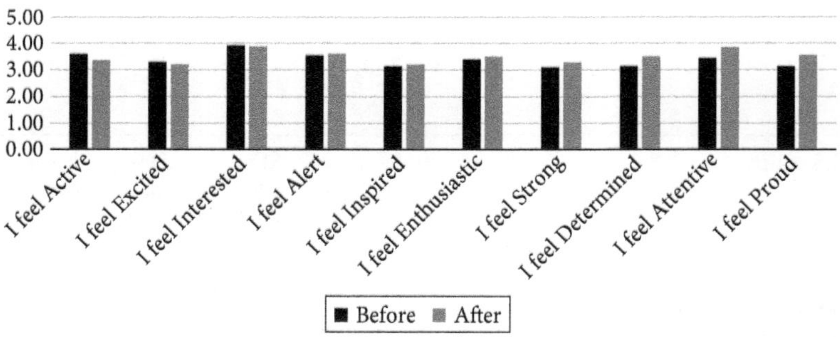

Appendix, Fig. 13 PANAS positive affect mean scores before and after visit to Priory Gardens.

Measure	P Value	SD Change	SD Percentage Change	Mean % Change
Enthusiastic	0.710	−0.27	−21.43	3.20
Ashamed	0.688	−0.59	−60.20	−18
Proud	0.263	−0.26	−18.18	13.91
Nervous	0.109	−0.44	−43.14	−19.64
Strong	0.332	−0.13	−10.00	6.14
Attentive	*0.043	−0.12	−9.68	12.70
Afraid	1.000	−0.11	−20.00	0.00
Determined	*0.019	−0.13	−9.63	12.17
Alert	0.648	−0.08	−6.96	2.31
Scared	0.125	−0.04	−7.69	9.76
Distressed	1.000	−0.01	−1.43	−2.08
Interested	0.813	−0.05	−4.42	−0.69
Inspired	0.845	−0.02	−1.53	2.61
Irritable	1.000	0.12	16.90	4.08
Upset	1.000	0.14	26.92	7.32
Guilty	0.219	0.24	44.44	14.29
Active	0.832	0.05	4.59	−6.06
Jittery	0.062	0.27	56.25	16.67
Excited	0.665	0.08	7.84	−2.48
Hostile	1.000	0.37	160.87	10.26

Appendix, Fig. 14 PANAS results before and after visiting Priory Gardens, including mean score percentage change, standard deviation change, standard deviation percentage change, and P value.

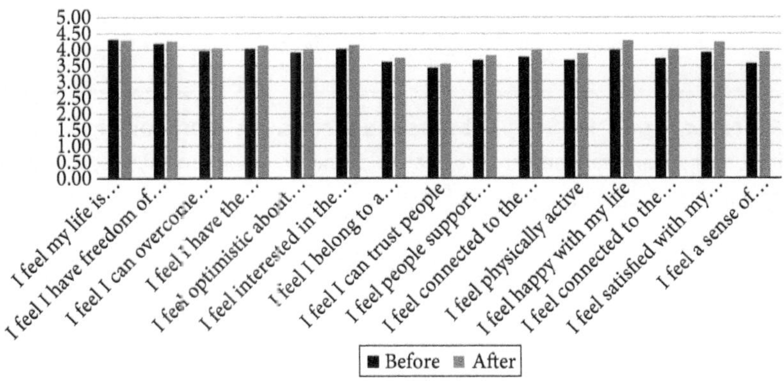

Appendix, Fig. 15 Modified Wellbeing Scale mean results before and after visiting Priory Gardens.

Measure	P Value	SD Change	SD Percentage Change	Mean Percentage Change
I feel I belong to a neighbourhood	0.629	−0.26	−20.31	3.76
I feel a sense of accomplishment	*0.013	−0.21	−19.63	10.69
I feel connected to the people around me	*0.039	−0.12	−12.37	8.03
I feel I can overcome challenges	1.000	−0.12	−11.43	2.05
I feel satisfied with my life	*0.013	−0.1	−10.10	8.33
I feel physically active	0.143	−0.11	−9.91	5.93
I feel people support each other	0.359	−0.1	−9.90	4.44
I feel connected to the local environment	0.096	−0.03	−3.23	5.76
I feel my life is worthwhile	1.000	−0.01	−1.18	−0.63
I feel I have freedom of choice	0.607	−0.01	−1.04	1.95
I feel I have the opportunity to learn new things	0.454	−0.01	−0.90	2.70
I feel happy with my life	*0.006	0	0.00	7.48
I feel I can trust people	0.581	0	0.00	3.97
I feel interested in the world around me	0.454	0.04	4.40	3.38
I feel optimistic about my future	0.434	0.06	6.59	2.78

Appendix, Fig. 16 MWS results before and after visiting Priory Gardens, including mean score percentage change, standard deviation change, standard deviation percentage change, and P value.

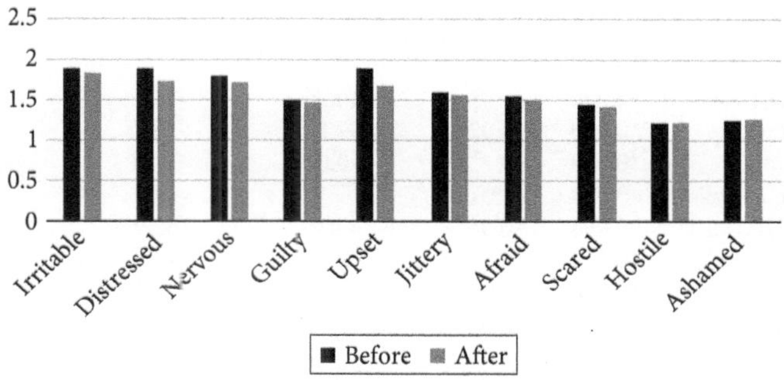

Appendix, Fig. 17 PANAS negative affects means scores before and after visiting Biltmore.

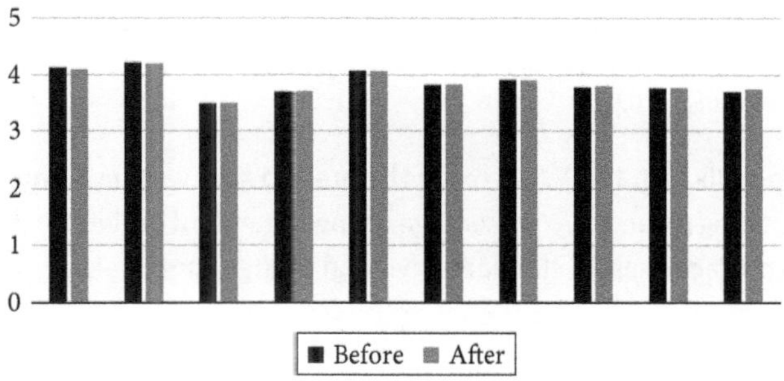

Appendix, Fig. 18 PANA Positive Affect Scale mean scores before and after visiting Biltmore.

Measure	P Value	SD Change	SD % Change	Mean % Change
Distressed	*0.000	−0.08	−10.00	−7.37
Irritable	*0.000	−0.02	−2.50	−5.57
Upset	*0.010	0	0.00	−3.20
Inspired	*0.047	−0.01	−1.16	1.50
Nervous	0.054	−0.04	−4.76	−3.19
Ashamed	0.140	0	0.00	1.44
Hostile	0.218	0.04	8.33	2.30
Strong	0.348	0.02	2.35	0.49
Proud	0.380	0.01	1.18	0.84
Guilty	0.448	−0.05	−6.85	−1.66
Excited	0.560	0.01	1.14	0.48
Afraid	0.640	−0.02	−2.86	−3.14
Jittery	0.655	0.02	2.82	−1.90
Scared	0.664	0	0.00	−0.86
Enthusiastic	0.690	−0.03	−3.61	0.18
Attentive	0.796	0.06	9.68	−0.44
Interested	0.902	0.01	1.56	−0.26
Alert	0.905	−0.02	−2.74	0.14
Active	0.906	0	0.00	0.11
Determined	0.908	0.01	1.22	−0.07

Appendix, Fig. 19 PANAS results before and after visiting Biltmore, including mean score percentage change, standard deviation percentage change, standard deviation change, and P value.

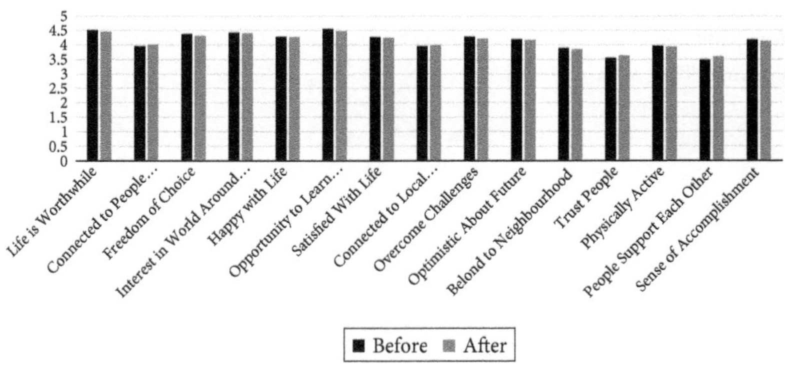

Appendix, Fig. 20 Modified Wellbeing Scale mean results before and after visiting Biltmore.

Measure	P Value	SD Change	SD % Change	Mean % Change
I feel people support each other	*0.000	0.01	1.25	3.31
I feel I can trust people	*0.000	0.03	3.90	2.71
I feel I have the opportunity to learn new things	*0.004	0.01	1.59	−1.40
I feel connected to the people around me	*0.013	−0.03	−3.90	1.44
I feel I have freedom of choice	*0.015	−0.01	−1.41	−1.33
I feel I can overcome challenges	*0.021	0.02	3.08	−1.33
I feel my life is worthwhile	0.086	0.04	6.35	−1.14
I feel physically active	0.128	0	0.00	−0.84
I feel a sense of accomplishment	0.177	−0.01	−1.37	−1.06
I feel connected to the local environment	0.219	−0.02	−2.50	0.98
I feel I belong to a neighbourhood	0.283	−0.03	−3.00	−1.04
I feel optimistic about my future	0.692	0.01	1.41	−0.17
I feel happy with my life	0.827	−0.01	−1.56	0.00
I feel interested in the world around me	1.000	−0.02	−3.03	−0.13
I feel satisfied with my life	1.000	−0.02	−2.94	−0.13

Appendix, Fig. 21 MWS results before and after visiting Biltmore Estate, including mean percentage changes, standard deviation change, SD percentage change, and significance.

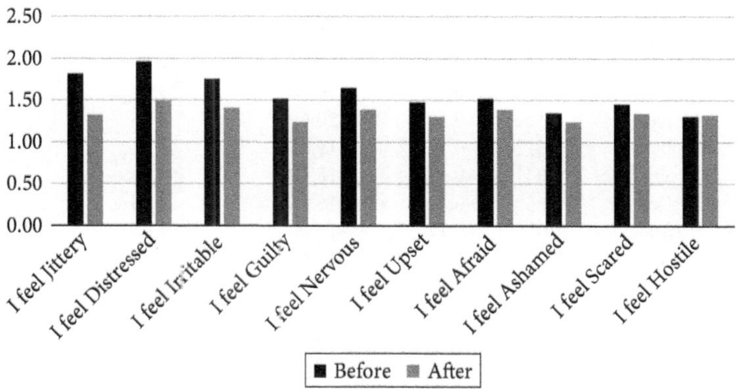

Appendix, Fig. 22 PANAS negative affects before and after visiting Nemours Estate.

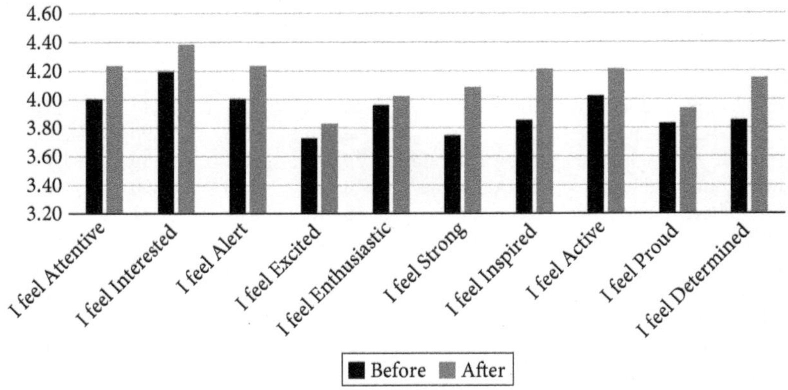

Appendix, Fig. 23 PANAS positive affect mean scores before and after visiting Nemours Estate.

Measure	P Value	SD Change	SD % Change	Mean % Change
Nervous	*0.049	−0.43	−38.74	−15.58
Irritable	*0.001	−0.36	−35.64	−19.51
Jittery	*0.000	−0.35	−32.41	−27.06
Determined	0.189	−0.3	−27.78	7.73
Strong	*0.013	−0.27	−27.27	9.09
Upset	0.180	−0.16	−21.33	−11.59
Distressed	*0.000	−0.19	−20.88	−23.91
Active	0.302	−0.18	−20.69	4.76
Guilty	*0.012	−0.15	−19.23	−18.31
Inspired	*0.041	−0.16	−18.18	9.39
Excited	0.556	−0.17	−16.04	2.86
Hostile	1.000	−0.12	−16.00	2.86
Alert	0.210	−0.11	−14.10	5.85
Interested	0.096	−0.08	−12.31	4.57
Afraid	0.146	−0.09	−10.84	−8.45
Scared	0.146	−0.08	−10.67	−7.35
Attentive	*0.049	−0.07	−10.45	5.85
Ashamed	0.227	−0.04	−6.25	−7.94
Enthusiastic	1.000	−0.01	−1.20	1.61
Proud	0.648	0.05	5.95	2.78

Appendix, Fig. 24 PANAS results before and after visiting Nemours Estate, including mean score percentage change, standard deviation change, standard deviation percentage change, and P value.

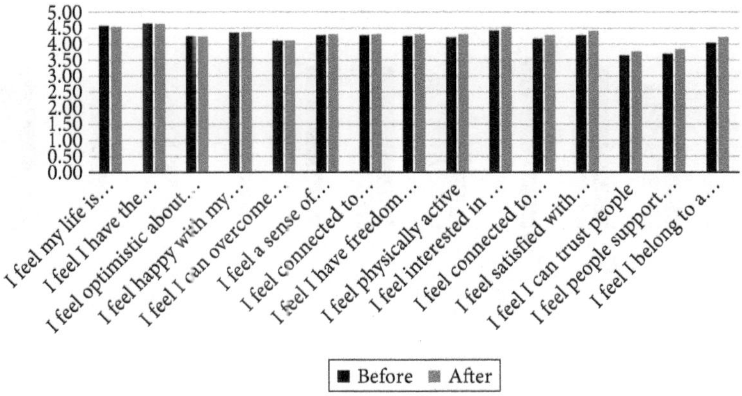

Appendix, Fig. 25 Modified Wellbeing Scale mean results before and after visiting Nemours Estate.

Measure	P Value	SD Change	SD % Change	Mean % Change
I feel satisfied with my life	0.180	−0.17	−23.94	3.50
I feel physically active	0.454	−0.18	−23.38	2.54
I feel interested in the world around me	0.210	−0.15	−23.08	2.90
I feel optimistic about my future	1.000	−0.14	−17.28	0.00
I feel I belong to a neighbourhood	0.057	−0.15	−17.24	4.76
I feel a sense of accomplishment	0.791	−0.12	−16.90	1.00
I feel my life is worthwhile	1.000	−0.1	−15.38	−0.47
I feel I have freedom of choice	0.607	−0.07	−9.21	1.51
I feel connected to the local environment	1.000	−0.05	−7.46	1.00
I feel connected to the people around me	0.210	−0.04	−5.80	3.08
I feel I have the opportunity to learn new things	1.000	0	0.00	0.00
I feel happy with my life	1.000	0.01	1.67	0.49
I feel I can trust people	0.388	0.05	6.58	3.51
I feel people support each other	0.210	0.1	14.49	4.05
I feel I can overcome challenges	0.481	0.14	17.50	0.52

Appendix, Fig. 26 Modified Wellbeing Scale results before and after visiting Nemours Estate, including mean score percentage change, standard deviation, SD percentage change, and significance.

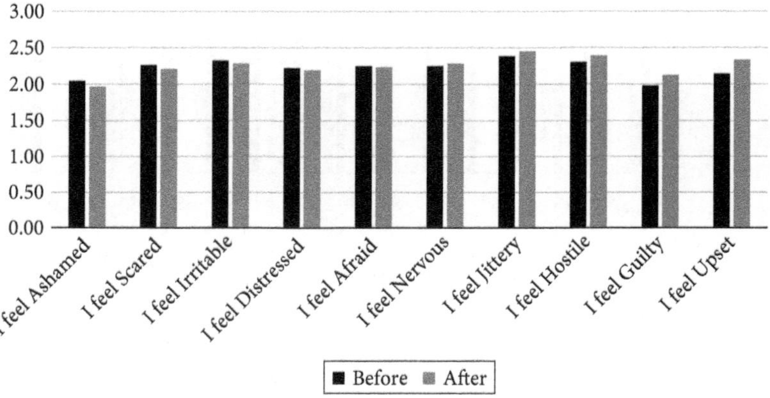

Appendix, Fig. 27 PANAS negative affects before and after visiting Bangalore Fort.

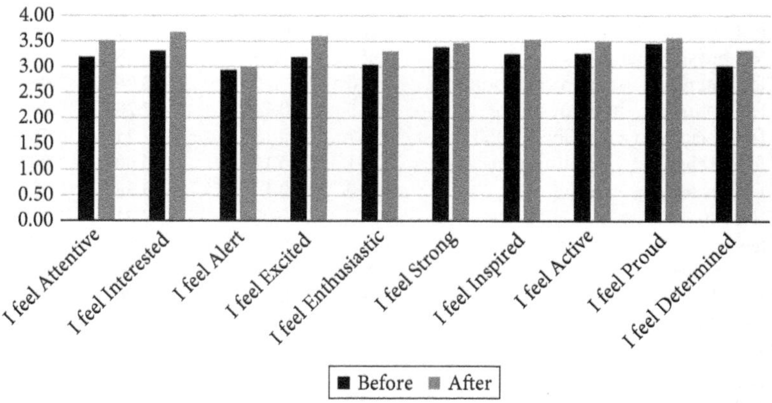

Appendix, Fig. 28 PANAS positive affect mean scores before and after visiting Bangalore Fort.

Measure	P Value	SD Change	SD % Change	Mean % Change
Excited	*0.002	–0.08	–7.14	13.03
Active	0.055	–0.05	–4.00	7.49
Attentive	*0.000	–0.01	–0.89	10.34
Determined	*0.001	0	0.00	10.12
Interested	*0.007	0.02	1.83	11.07
Proud	0.127	0.05	4.85	3.18
Enthusiastic	*0.046	0.07	5.98	8.84
Inspired	*0.006	0.06	6.32	9.02
Ashamed	0.728	–0.05	–4.13	–3.59
Scared	0.664	–0.03	–2.36	–2.16
Alert	0.371	0.12	10.43	2.50
Afraid	1.000	0	0.00	–0.54
Strong	0.291	0.13	11.61	2.53
Irritable	0.310	0.03	2.42	–1.58
Guilty	0.151	0.03	2.52	7.41
Upset	0.216	0.07	5.93	9.14
Distressed	0.719	0.07	6.14	–1.10
Jittery	0.441	0.1	8.13	3.08
Nervous	0.864	0.11	9.02	1.63
Hostile	0.710	0.22	18.33	4.26

Appendix, Fig. 29 PANAS results before and after visiting Bangalore Fort, including mean score percentage change, SD change, standard deviation percentage change, and P value.

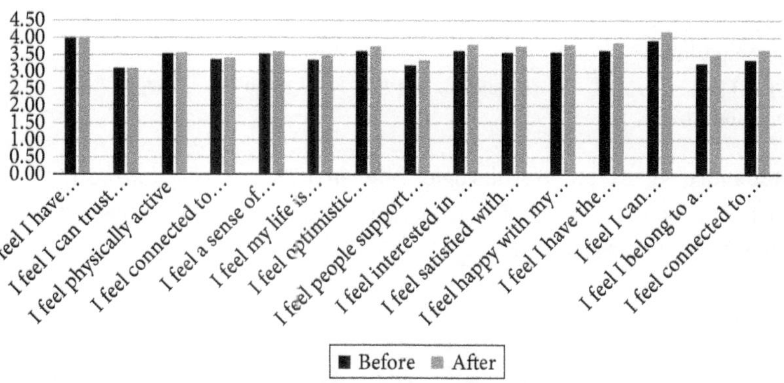

Appendix, Fig. 30 Modified Wellbeing Scale mean results before and after visiting Bangalore Fort.

Measure	P Value	SD Change	SD % Change	Mean % Change
I feel I can overcome challenges	*0.018	−0.17	−15.89	6.88
I feel optimistic about my future	0.302	−0.1	−9.17	4.42
I feel happy with my life	*0.002	−0.04	−3.60	6.51
I feel I have the opportunity to learn new things	*0.010	−0.02	−1.79	6.76
I feel physically active	0.770	0.01	0.86	1.04
I feel satisfied with my life	*0.045	0.02	1.92	5.84
I feel connected to the people around me	*0.012	0.03	2.97	9.16
I feel people support each other	0.074	0.04	3.23	5.38
I feel connected to the local environment	0.280	0.04	3.39	1.45
I feel I can trust people	0.651	0.05	3.65	0.40
I feel interested in the world around me	*0.008	0.05	4.72	5.42
I feel my life is worthwhile	0.118	0.07	5.83	4.03
I feel I have freedom of choice	1.000	0.08	8.00	0.31
I feel a sense of accomplishment	0.312	0.12	12.24	2.08
I feel I belong to a neighbourhood	*0.004	0.17	15.89	8.81

Appendix, Fig. 31 MWS results before and after visiting Bangalore Fort, including mean score percentage change, standard deviation percentage change, SD change, and P value.

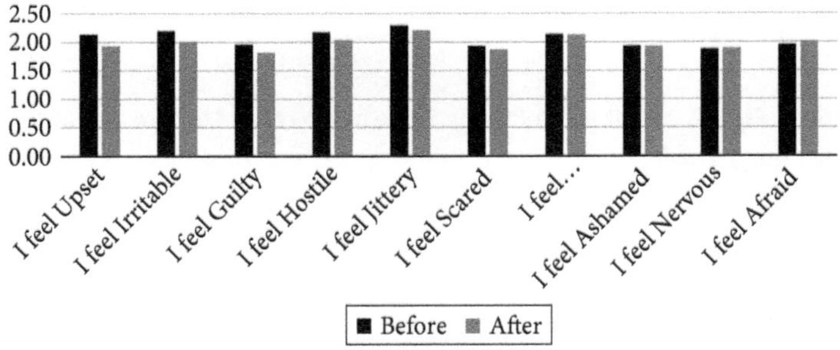

Appendix, Fig. 32 PANAS negative affects before and after visiting TSSP.

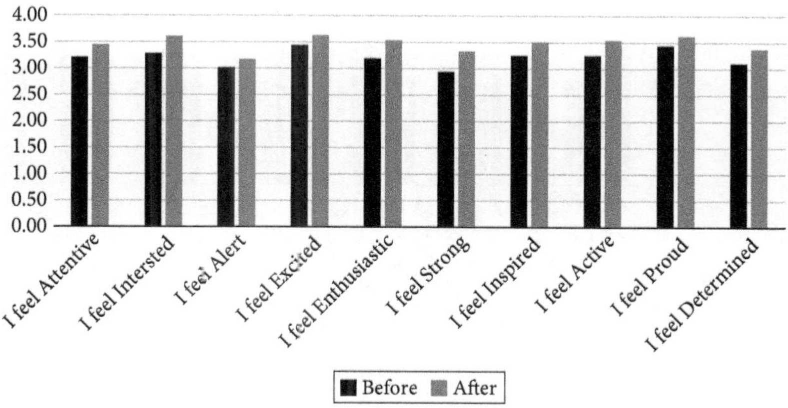

Appendix, Fig. 33 PANAS positive affects before and after visiting TSSP.

Measure	P Value	SD Change	SD % Change	Mean % Change
Active	*0.005	−0.07	−5.34	9.32
Excited	*0.050	−0.05	−4.46	5.78
Enthusiastic	*0.002	−0.01	−0.86	11.15
Inspired	*0.007	0.01	0.93	8.04
Determined	*0.004	0.04	3.60	9.09
Attentive	*0.005	0.04	3.81	7.49
Interested	*0.000	0.06	5.26	10.19
Alert	0.161	0.09	7.83	5.56
Proud	*0.015	0.12	11.54	5.47
Upset	0.090	−0.10	−8.06	−9.31
Jittery	0.728	−0.02	−1.59	−3.65
Hostile	0.511	0.01	0.80	−5.80
Strong	*0.000	0.15	12.93	13.52
Irritable	0.091	0.03	2.44	−8.57
Guilty	0.134	0.03	2.54	−6.95
Distressed	0.522	0.07	5.98	0
Scared	0.584	0.08	7.02	−2.72
Nervous	0.868	0.10	9.17	1.12
Ashamed	0.377	0.14	12.07	0
Afraid	0.377	0.21	19.27	3.76

Appendix, Fig. 34 PANAS results before and after visiting TSSP including mean score percentage change, SD mean change, standard deviation percentage change, and P values.

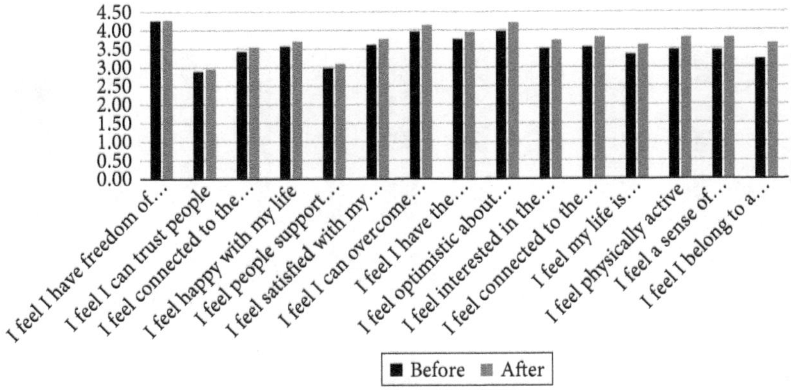

Appendix, Fig. 35 MWS mean scores before and after visiting TSSP.

Measure	P Value	SD Change	Standard Deviation % Change	Mean % Change
I feel connected to the people around me	*0.033	−0.97	−89.91	8.15
I feel I belong to a neighbourhood	*0.000	−0.10	−8.13	13.31
I feel my life is worthwhile	*0.003	−0.01	−0.87	7.99
I feel optimistic about my future	*0.017	0.00	0.00	6.05
I feel I can overcome challenges	*0.049	0.00	0.00	4.76
I feel I have freedom of choice	0.349	0.00	0.00	0.74
I feel satisfied with my life	*0.041	0.01	0.88	4.64
I feel interested in the world around me	*0.002	0.03	2.78	6.57
I feel I can trust people	0.391	0.07	5.43	2.54
I feel connected to the local environment	0.086	0.07	6.19	3.98
I feel physically active	*0.003	0.07	6.60	9.58
I feel a sense of accomplishment	*0.001	0.07	6.67	10.30
I feel people support each other	0.145	0.08	7.41	4.21
I feel happy with my life	0.050	0.09	7.96	4.11
I feel I have the opportunity to learn new things	*0.018	0.11	11.34	5.28

Appendix, Fig. 36 MWS results before and after visiting TSSP, including mean score percentage change, SD mean change, standard deviation percentage change, and P values.

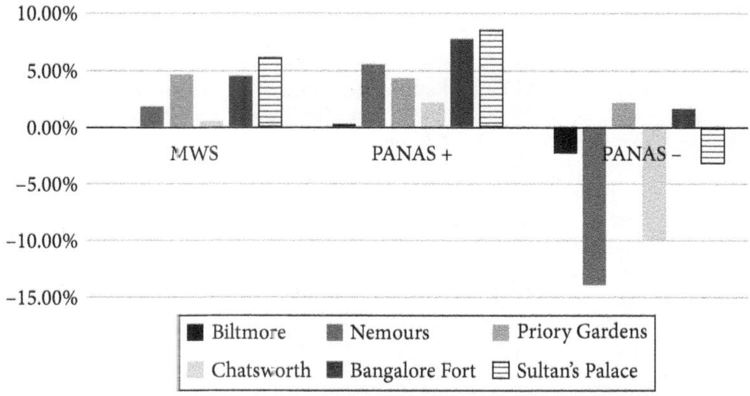

Appendix, Fig. 37 Mean average percentage change PANAS positive, negative, and MWS for all case studies.

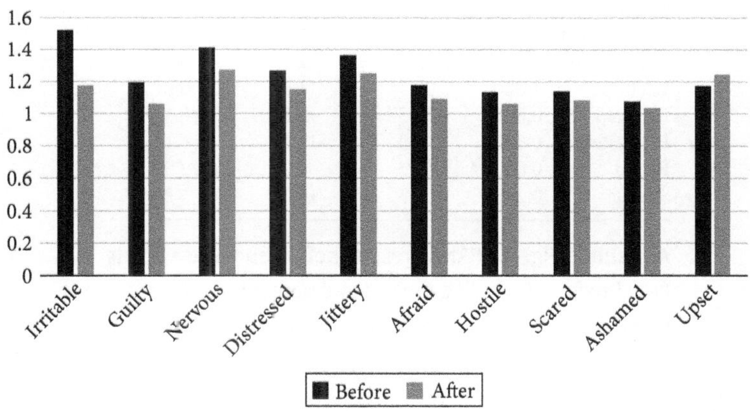

Appendix, Fig. 38 PANAS negative affect mean scores before and after visiting IWM Duxford.

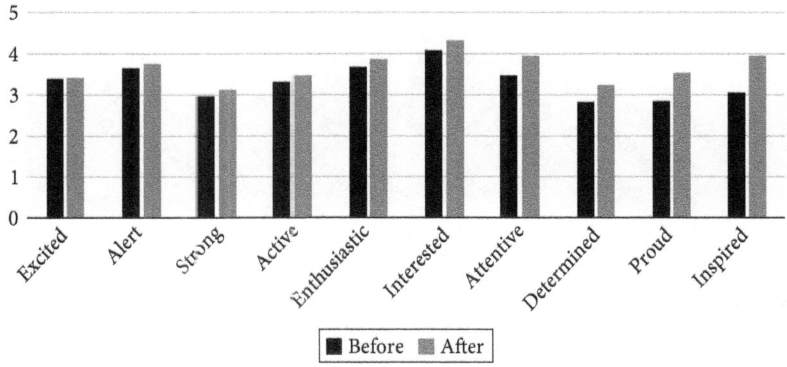

Appendix, Fig. 39 PANAS: Positive Affect Scale mean scores before and after visiting IWM Duxford.

Measure	P Value	Standard Deviation Change	Standard Deviation % Change	Mean Score % Change
Afraid	0.109	−0.12	−26.67	−6.25
Ashamed	0.688	−0.10	−33.33	−2.91
Distressed	0.019	−0.16	−28.07	−9.09
Guilty	0.039	−0.38	−61.29	−10.53
Hostile	0.774	−0.19	−40.43	−5.56
Irritable	0.000*	−0.38	−44.19	−23.29
Jittery	0.043*	0.04	5.71	−8.40
Nervous	0.015*	0.00	0.00	−9.63
Scared	0.344	−0.09	−22.50	−4.59
Upset	0.388	0.12	25.00	6.25
Active	0.212	0.02	1.90	5.70
Alert	0.590	0.00	0.00	3.43
Attentive	0.000*	−0.06	−6.06	13.81
Determined	0.000*	0.11	10.00	15.13
Enthusiastic	0.120	0.02	2.02	5.98
Excited	1.000	0.12	11.65	0.61
Inspired	0.000*	−0.03	−2.88	30.14
Interested	0.005*	−0.09	−10.23	6.14
Proud	0.000*	−0.03	−2.36	24.09
Strong	0.243	0.17	16.50	5.26

Appendix, Fig. 40 PANAS results before and after visiting IWM Duxford, including mean score percentage change, standard deviation percentage change, and P values.

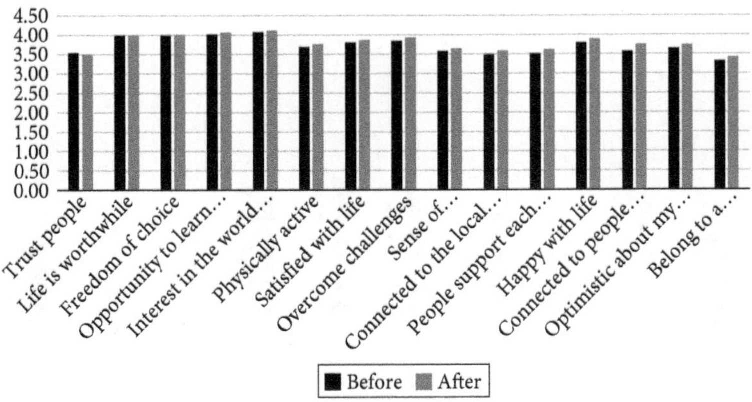

Appendix, Fig. 41 MWS mean results before and after visiting IWM Duxford.

Measure	P Value	Standard Deviation Change	SD % Change	Mean % Change
I feel a sense of accomplishment	0.749	0.01	1.14	2.65
I feel connected to the local environment	0.542	−0.07	−7.07	2.70
I feel connected to the people around me	0.127	−0.06	−6.82	5.88
I feel happy with my life	*0.043	0.00	0.00	3.32
I feel I belong to a neighbourhood	0.324	0.00	0.00	3.49
I feel I can overcome challenges	0.607	−0.12	−13.95	2.45
I feel I can trust people	0.700	0.04	4.94	−0.30
I feel I have freedom of choice	1.000	−0.14	−14.74	0.52
I feel I have the opportunity to learn new things	0.868	−0.03	−3.95	1.57
I feel interested in the world around me	1.000	−0.03	−4.00	1.03
I feel my life is worthwhile	0.571	−0.03	−3.12	0.00
I feel optimistic about my future	0.391	0.00	0.00	2.01
I feel people support each other	0.296	−0.01	−1.12	3.28
I feel physically active	1.000	−0.02	−2.13	1.70
I feel satisfied with my life	0.089	0.06	8.00	2.20

Appendix, Fig. 42 MWS results before and after visiting IWM Duxford, including mean score percentage change, SD mean score, standard deviation percentage change, and P values.

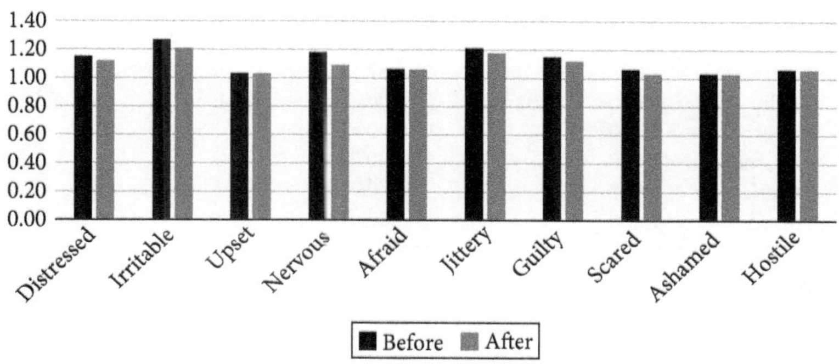

Appendix, Fig. 43 PANAS negative affect mean scores before and after visiting Dulwich Picture Gallery.

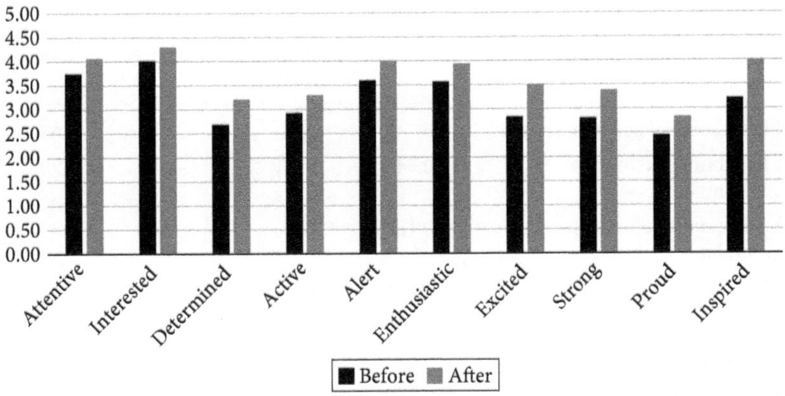

Appendix, Fig. 44 PANAS positive affect mean scores before and after visiting Dulwich Picture Gallery.

Measure	P Value	SD Change	SD % Change	Mean Score % Change
Alert	*0.049	−0.31	−27.43	11.48
Attentive	0.143	−0.36	−34.29	8.66
Inspired	*0.000	−0.19	−18.81	−4.65
Excited	*0.000	−0.16	−12.40	23.96
Active	0.143	−0.08	−7.21	13.13
Interested	*0.006	−0.06	−8.11	7.35
Enthusiastic	*0.027	−0.04	−4.30	10.74
Proud	*0.065	0.08	6.35	15.66
Strong	*0.035	0.03	2.80	21.05
Scared	1.000	−0.07	−29.17	−2.78
Determined	*0.013	0.05	4.00	19.78
Irritable	0.688	−0.16	−28.07	−4.65
Nervous	0.375	−0.08	−17.39	−7.50
Jittery	0.625	−0.07	−11.86	−2.44
Distressed	1.000	−0.03	−6.82	−2.56
Guilty	1.000	−0.02	−4.00	−2.56
Upset	1.000	0	0.00	0
Ashamed	1.000	0	0.00	0
Hostile	1.000	0	0.00	0
Afraid	1.000	0	0.00	0

Appendix, Fig. 45 PANAS results before and after visiting DPG, including mean score percentage change, standard deviation percentage change, SD mean score change, and P value.

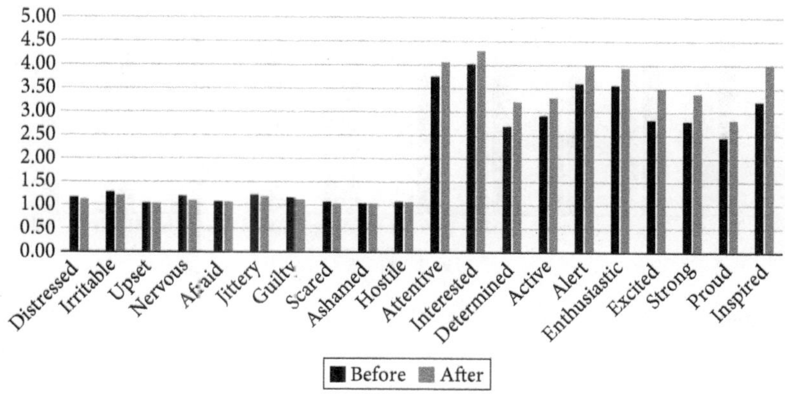

Appendix, Fig. 46 MWS mean scores before and after visiting Dulwich Picture Gallery.

Measure	P Value	SD Change	SD % Change	Mean % Change
I feel interested in the world around me	*0.004	−0.2	−26.32	9.49
I feel I have the opportunity to learn new things	*0.000	−0.19	−25.33	10.79
I feel I can overcome challenges	*0.012	−0.17	−20.99	8.53
I feel I have freedom of choice	0.267	−0.15	−19.23	4.58
I feel a sense of accomplishment	*0.000	−0.15	−16.67	17.80
I feel my life is worthwhile	*0.039	−0.14	−16.67	6.82
I feel satisfied with my life	*0.004	−0.06	−8.22	7.58
I feel I can trust people	0.581	−0.1	−12.99	2.42
I feel happy with my life	*0.006	−0.04	−5.48	8.33
I feel I belong to a neighbourhood	0.549	−0.12	−12.90	5.04
I feel connected to the local environment	0.146	−0.08	−9.76	7.38
I feel physically active	1.000	−0.08	−8.08	0.76
I feel connected to the people around me	0.210	0	0.00	9.49
I feel people support each other	0.607	0.03	4.35	1.61
I feel optimistic about my future	0.109	0.08	10.67	5.83

Appendix, Fig. 47 MWS before and after visiting DPG, including mean score percentage change, standard deviation percentage change, mean score change, and P value.

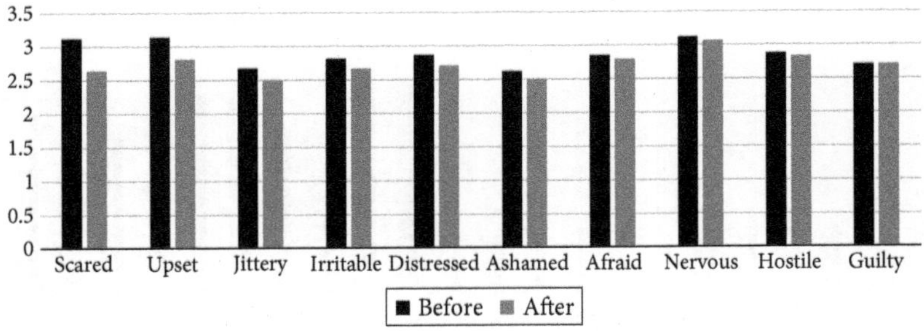

Appendix, Fig. 48 PANAS negative affect mean scores before and after visiting CGMAG.

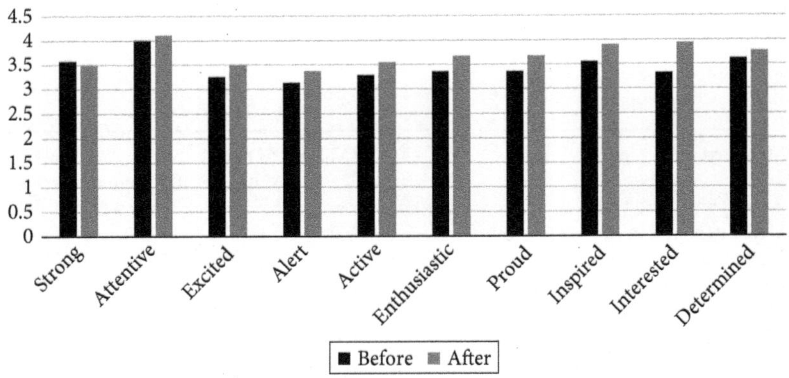

Appendix, Fig. 49 PANAS positive affect mean scores before and after visiting CGMAG.

Measure	P Value	SD Change	SD % Change	Mean % Change
Attentive	0.324	−0.02	−2.47	3.23
Active	*0.037	−0.02	−2.30	8.30
Interested	*0.000	0.02	2.41	18.97
Irritable	0.429	−0.13	−11.61	−5.08
Inspired	*0.003	0.04	4.55	10.08
Upset	*0.010	−0.04	−4.04	−10.05
Enthusiastic	*0.037	0.05	6.02	9.83
Determined	0.074	0.09	11.25	4.35
Afraid	0.582	−0.04	−3.92	−2.01
Alert	0.117	0.11	12.09	7.08
Proud	*0.007	0.10	12.35	9.83
Ashamed	0.127	−0.04	−3.54	−4.37
Excited	*0.012	0.12	15.19	7.49
Scared	*0.027	−0.02	−1.94	−15.14
Jittery	*0.040	0.03	3.03	−6.95
Nervous	0.662	0.04	4.76	−1.83
Distressed	0.499	0.09	7.83	−5.00
Strong	0.877	0.17	20.48	−2.01
Hostile	1.000	0.13	13.54	−1.49
Guilty	0.939	0.24	23.76	0.53

Appendix, Fig. 50 PANAS positive and negative affect results before and after visiting CGMAG, including mean score percentage change, standard deviation percentage change, SD mean score change, and P value.

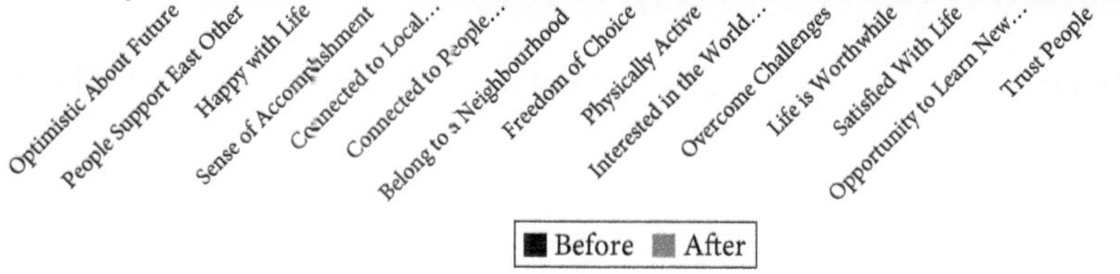

Appendix, Fig. 51 MWS mean scores before and after visiting CGMAG.

Measure	P Value	SD Change	SD % Change	Mean % Change Scores
I feel interested in the world around me	0.123	−0.18	−18.00	5.81
I feel connected to the people around me	0.607	−0.16	−15.53	4.44
I feel connected to the local environment	0.291	−0.15	−15.00	4.26
I feel I belong to a neighbourhood	0.176	−0.12	−13.95	4.44
I feel I can overcome challenges	0.417	−0.12	−12.24	6.72
I feel satisfied with my life	*0.009	−0.11	−11.58	11.91
I feel I have freedom of choice	0.337	−0.11	−11.22	4.74
I feel I have the opportunity to learn new things	*0.000	−0.08	−8.60	12.77
I feel a sense of accomplishment	0.635	−0.08	−8.33	4.05
I feel physically active	0.175	−0.07	−7.45	4.74
I feel people support each other	0.216	−0.05	−5.88	3.25
I feel I can trust people	*0.009	−0.06	−5.71	12.77
I feel happy with my life	0.735	−0.01	−1.14	3.64
I feel optimistic about my future	0.511	−0.01	−1.01	3.25
I feel my life is worthwhile	*0.009	0.03	3.41	7.66

Appendix, Fig. 52 MWS results before and after visiting CGMAG, including mean score percentage change, standard deviation percentage change, and P values.

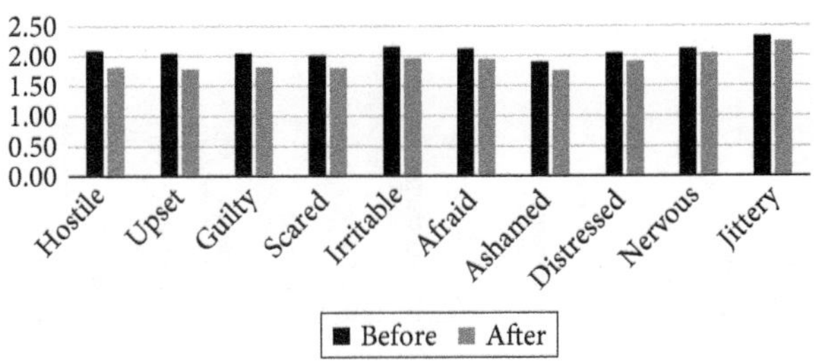

Appendix, Fig. 53 PANAS negative affect mean scores before and after visiting VAG.

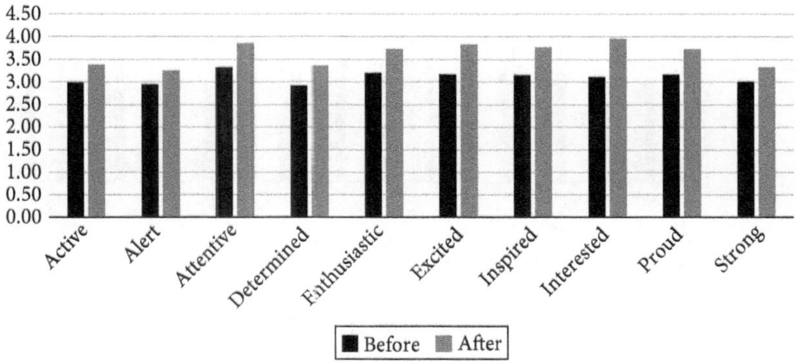

Appendix, Fig. 54 PANAS positive affect mean scores before and after visiting VAG.

Measure	P Value	SD Change	Standard Deviation % Change	Mean Score % Change
Jittery	0.824	−0.09	−7.20	−3.42
Guilty	*0.043	−0.06	−5.22	−10.94
Hostile	*0.031	0.00	0.00	−12.98
Scared	0.067	0.01	0.86	−10.32
Irritable	0.100	0.04	3.54	−8.89
Nervous	0.124	0.05	4.35	−3.76
Afraid	0.063	0.06	5.08	−8.27
Ashamed	0.361	0.07	6.67	−7.66
Distressed	0.307	0.14	12.17	−6.25
Upset	*0.031	0.15	14.29	−12.50
Interested	*0.000	−0.03	−2.40	27.04
Attentive	*0.000	0.01	0.82	16.27
Excited	*0.001	0.02	1.71	21.11
Strong	*0.029	0.03	2.42	11.11
Determined	*0.000	0.07	5.60	15.58
Active	*0.001	0.13	10.08	13.30
Alert	0.188	0.14	10.22	10.81
Inspired	*0.000	0.14	12.39	19.70
Enthusiastic	*0.001	0.18	17.48	16.92
Proud	*0.000	0.19	17.76	18.09

Appendix, Fig. 55 PANAS positive and negative affect results before and after visiting VAG, including mean score percentage change, standard deviation percentage change, SD mean score, and P value.

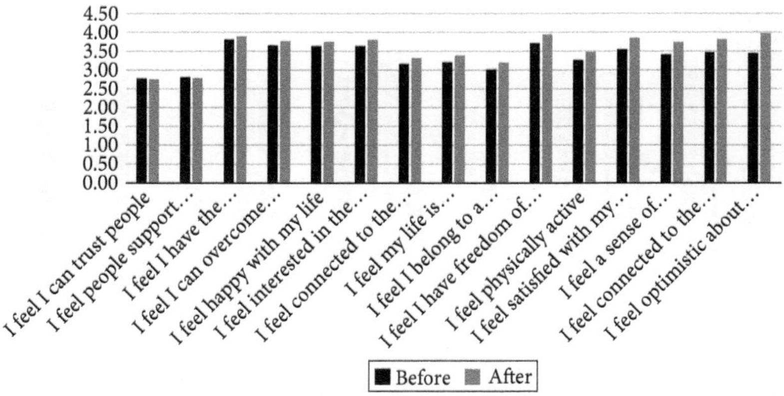

Appendix, Fig. 56 MWS mean scores before and after visiting VAG.

Measure	P Value	SD Change	SD Mean % Change	Mean Scores % Change
I feel optimistic about my future	*0.000	0.02	1.61	15.74
I feel satisfied with my life	*0.001	0	0	8.52
I feel a sense of accomplishment	*0.001	0.14	11.97	9.81
I feel connected to the people around me	*0.002	0	0	10.09
I feel physically active	*0.026	0.09	6.98	6.83
I feel my life is worthwhile	*0.031	-0.01	-0.78	5.56
I feel I have freedom of choice	*0.045	0.08	6.5	6.44
I feel interested in the world around me	0.064	-0.01	-0.79	4.82
I feel I belong to a neighbourhood	0.078	0.11	8.87	6.35
I feel connected to the local environment	0.100	0.07	5.56	5.97
I feel I can overcome challenges	0.152	0.08	6.84	3.49
I feel happy with my life	0.248	0.1	8.06	3.51
I feel I have the opportunity to learn new things	0.327	0.02	1.54	2.51
I feel I can trust people	0.458	0.06	4.08	-0.57
I feel people support each other	0.839	0.14	10.29	-0.57

Appendix, Fig. 57 MWS results before and after visiting VAG, including mean score percentage change, standard deviation percentage change, SD mean score change, and P values.

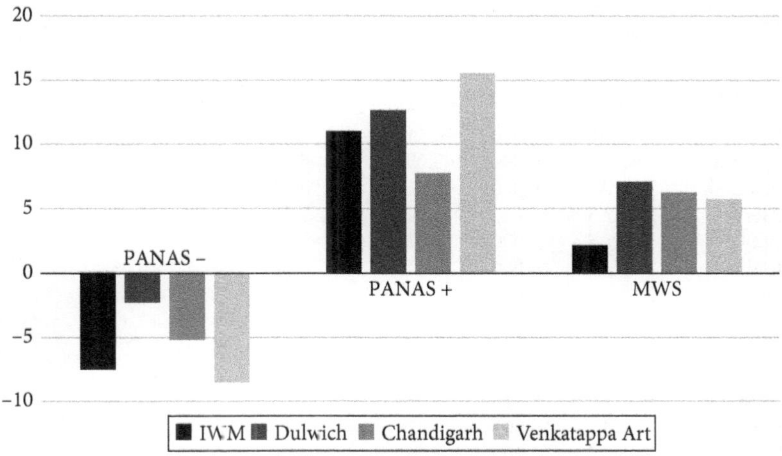

Appendix, Fig. 58 Mean average percentage change in PANAS and MWS for all museums and art gallery case study sites.

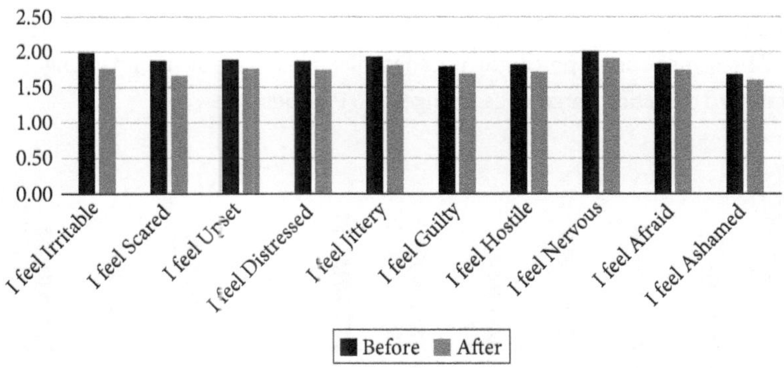

Appendix, Fig. 59 PANAS negative mean scores for museums and art galleries visitors.

Measure	Survey	P Value	SD Change	SD Percentage Change	Mean Percentage Change
Scared	PANAS –	*0.001	–0.14	–11.97	–10.63
Interested	PANAS +	*0.000	–0.11	–10.58	13.72
Attentive	PANAS +	*0.000	–0.09	–8.65	10.55
Inspired	PANAS +	*0.000	0	0.00	21.13
Alert	PANAS +	*0.010	0.02	1.77	7.20
Active	PANAS +	*0.000	0.02	1.85	9.01
Enthusiastic	PANAS +	*0.000	0.04	4.12	10.03
Determined	PANAS +	*0.000	0.05	4.39	12.41
Proud	PANAS +	*0.000	0.06	5.22	17.47
Irritable	PANAS –	*0.000	–0.09	–7.96	–11.13
Upset	PANAS –	*0.006	–0.06	–5.26	–6.28
Ashamed	PANAS –	*0.043	–0.03	–2.91	–4.55
Jittery	PANAS –	*0.003	–0.02	–1.82	–5.94
Excited	PANAS +	*0.000	0.06	5.71	9.91
Strong	PANAS +	*0.010	0.09	8.41	6.23
Afraid	PANAS –	*0.030	–0.02	–1.82	–4.58
Hostile	PANAS –	0.120	0	0.00	–5.46
Distressed	PANAS –	*0.020	0.02	1.77	–6.15
Guilty	PANAS –	*0.036	0.02	1.85	–5.53
Nervous	PANAS –	*0.006	0.03	2.65	–4.75

Appendix, Fig. 60 PANAS results before and after visiting museums and art galleries: mean score percentage change, mean standard deviation change, standard deviation percentage change, and P values.

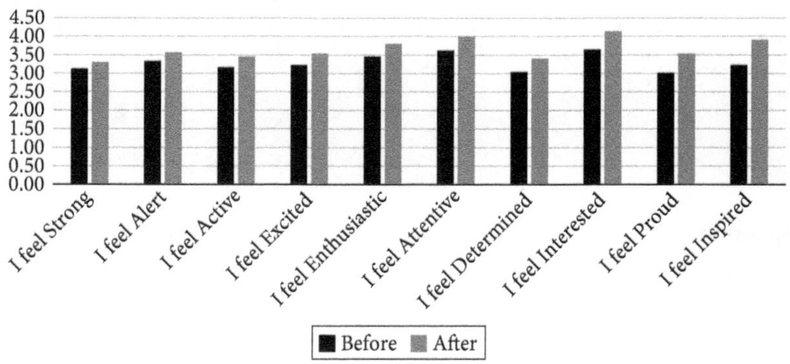

Appendix, Fig. 61 PANAS positive mean scores for museum and art gallery visitors.

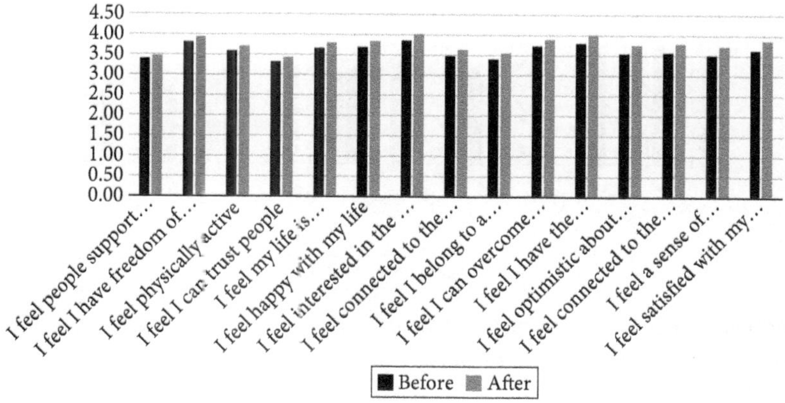

Appendix, Fig. 62 MWS mean scores for museum and art gallery visitors.

Measure	Sig. level	SD Change	SD Percentage Change	Mean MWS Percentage
I feel interested in the world around me	*0.003	−0.08	−8.16	4.24
I feel I have freedom of choice	*0.059	−0.07	−6.86	3.50
I feel I can overcome challenges	*0.009	−0.06	−6.19	4.60
I feel connected to the people around me	*0.001	−0.06	−5.77	6.41
I feel I have the opportunity to learn new things	*0.000	−0.05	−5.05	5.72
I feel connected to the local environment	*0.014	−0.04	−3.81	4.49
I feel satisfied with my life	*0.000	−0.03	−3.16	6.82
I feel my life is worthwhile	*0.001	−0.03	−2.83	4.07
I feel physically active	*0.025	−0.01	−0.95	3.51
I feel I belong to a neighbourhood	*0.010	0	0.00	4.59
I feel optimistic about my future	*0.000	0.01	0.99	6.02
I feel a sense of accomplishment	*0.000	0.01	1.02	6.64
I feel I can trust people	*0.011	0.03	2.70	3.56
I feel people support each other	0.078	0.04	3.81	2.36
I feel happy with my life	*0.003	0.04	4.30	4.13

Appendix, Fig. 63 MWS results before and after visiting museums and art galleries: mean score percentage change, standard deviation change, standard deviation percentage change, and P values.

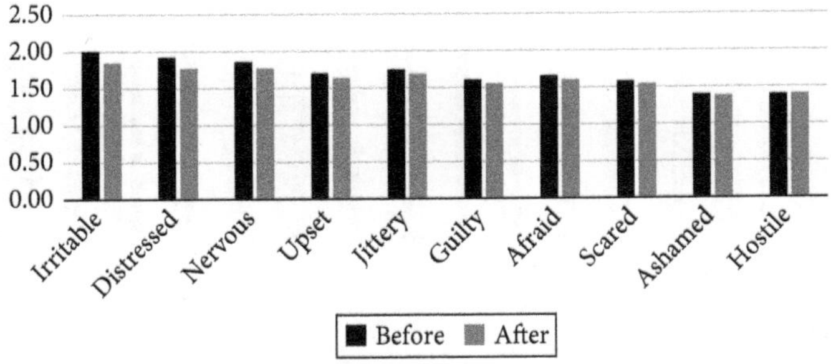

Appendix, Fig. 64 PANAS negative mean scores before and after visiting heritage and archaeological sites.

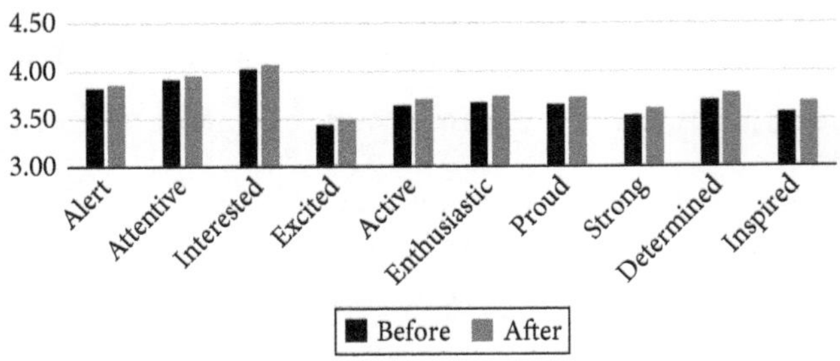

Appendix, Fig. 65 PANAS positive mean scores before and after visiting heritage and archaeological sites.

Measure	Survey	P Value	SD Change	SD % Change	Mean % Change
Determined	PANAS +	*0.001	−0.0449	−4.48	2.22
Guilty	PANAS −	0.063	−0.0324	−3.76	-2.95
Active	PANAS +	*0.012	−0.0437	−4.18	1.92
Enthusiastic	PANAS +	*0.012	−0.0371	−3.92	1.96
Irritable	PANAS −	*0.000	−0.0346	−3.74	-7.64
Distressed	PANAS −	*0.000	−0.0315	−3.58	-7.55
Nervous	PANAS −	*0.002	−0.0298	−3.18	-4.42
Interested	PANAS +	*0.036	−0.0253	−3.02	1.24
Inspired	PANAS +	*0.000	−0.0164	−1.73	3.44
Jittery	PANAS −	0.060	−0.0262	−2.74	-3.13
Alert	PANAS +	0.127	−0.0116	−1.25	1.12
Attentive	PANAS +	*0.038	−0.0111	−1.34	1.18
Scared	PANAS −	0.115	−0.0124	−1.50	-1.85
Ashamed	PANAS −	0.901	−0.0083	−1.08	-0.52
Proud	PANAS +	*0.003	−0.0064	−0.67	2.14
Upset	PANAS −	*0.001	−0.0043	−0.51	-3.65
Excited	PANAS +	*0.025	−0.0022	−0.23	1.88
Strong	PANAS +	*0.001	0.0197	2.03	2.19
Afraid	PANAS −	*0.016	0.0183	2.16	-2.87
Hostile	PANAS −	0.701	0.0406	5.26	0.85

Appendix, Fig. 66 PANAS results before and after visiting museums and art galleries: mean score percentage change, mean score standard deviation change, standard deviation percentage change, and P values.

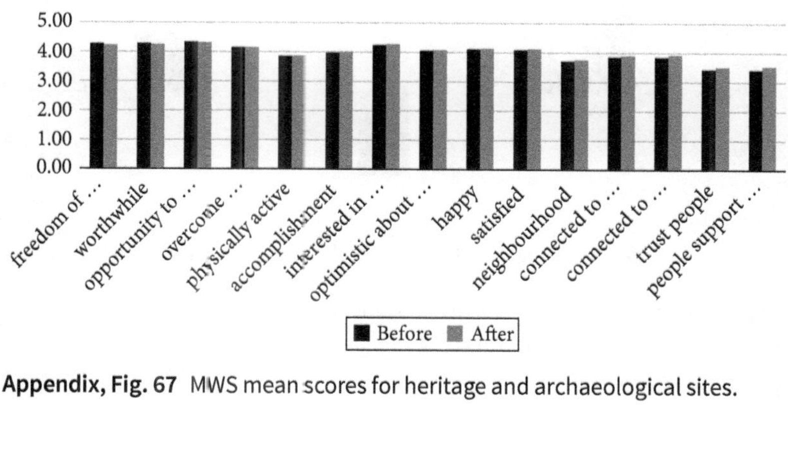

Appendix, Fig. 67 MWS mean scores for heritage and archaeological sites.

Measure	P Value	SD Change	SD % Change	Mean % Change
I feel I belong to a neighbourhood	*0.022	−0.0521	−4.82	1.38
I feel satisfied with my life	*0.021	−0.038	−4.64	1.07
I feel connected to the people around me	*0.000	−0.0342	−4.00	2.42
I feel connected to the local environment	*0.014	−0.0136	−1.52	1.40
I feel interested in the world around me	0.053	−0.032	−3.95	0.88
I feel a sense of accomplishment	0.083	−0.0335	−3.90	0.87
I feel optimistic about my future	0.241	−0.0223	−2.68	0.90
I feel I have the opportunity to learn new things	0.607	−0.0189	−2.31	−0.08
I feel happy with my life	*0.014	−0.011	−1.39	0.93
I feel physically active	0.408	−0.0196	−2.01	0.81
I feel my life is worthwhile	0.629	−0.0124	−1.42	−0.43
I feel I can overcome challenges	1.000	−0.0072	−0.94	0.18
I feel I have freedom of choice	0.164	−0.001	−0.12	−0.81
I feel people support each other	*0.000	0.0181	2.04	3.60
I feel I can trust people	*0.000	0.0389	4.25	2.60

Appendix, Fig. 68 MWS results before and after visiting heritage and archaeological sites: mean score percentage change, standard deviation percentage change, mean score standard deviation change, and P value.

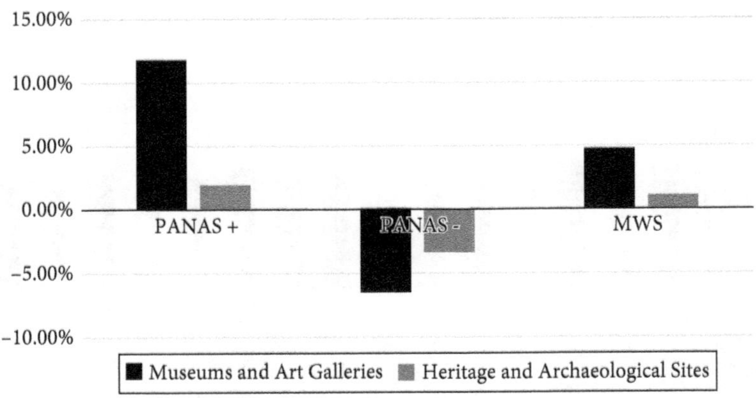

Appendix, Fig. 69 Comparison of MWS and PANAS of percentage change in mean scores from heritage and archaeological sites, and museums and art galleries.

Measure	Survey	P-Value	SD Change	SD % Change	Mean % Change
Irritable	PANAS –	*0.000	–0.21	–21.88	–17.21
Ashamed	PANAS –	0.532	–0.11	–18.03	-3.71
Guilty	PANAS –	*0.007	–0.14	–17.28	-8.03
Nervous	PANAS –	*0.000	–0.1	–10.87	-11.91
Attentive	PANAS +	*0.002	–0.1	–10.53	5.78
Distressed	PANAS –	*0.000	–0.08	–10.81	–10.85
Alert	PANAS +	0.078	–0.05	–5.15	3.51
Proud	PANAS +	*0.000	–0.06	–4.88	12.90
Active	PANAS +	*0.003	–0.05	–4.55	5.03
Interested	PANAS +	*0.028	–0.02	–2.33	2.29
Determined	PANAS +	*0.000	–0.02	–1.74	16.14
Scared	PANAS –	0.071	–0.06	–8.33	–3.83
Enthusiastic	PANAS +	*0.006	–0.01	–1.04	4.87
Inspired	PANAS +	*0.000	0.03	2.88	16.14
Afraid	PANAS –	*0.016	–0.05	–6.85	–5.94
Strong	PANAS +	*0.012	0.07	6.80	5.31
Excited	PANAS +	0.132	0.07	6.80	2.82
Hostile	PANAS –	0.281	–0.03	–5.88	–2.91
Upset	PANAS –	0.201	–0.03	–4.11	–3.19
Jittery	PANAS –	*0.001	0	0.00	–5.80

Appendix, Fig. 70 PANAS results before and after visiting UK heritage places: mean score percentage change, mean standard deviation change, standard deviation percentage change, and P value.

Measure	P Value	SD Change	SD Percentage Change	Mean Percentage Change
I feel interested in the world around me	0.271	−0.08	−10.13	1.54
I feel I can overcome challenges	*0.042	−0.07	−8.43	2.59
I feel physically active	0.137	−0.07	−6.93	2.72
I feel connected to the local environment	0.057	−0.06	−6.59	2.90
I feel connected to the people around me	0.085	−0.05	−5.88	2.97
I feel I belong to a neighbourhood	0.160	−0.05	−4.67	3.07
I feel I have freedom of choice	0.777	−0.04	−4.60	0.45
I feel my life is worthwhile	0.133	−0.04	−4.44	-0.96
I feel people support each other	*0.030	−0.03	−3.49	3.26
I feel a sense of accomplishment	*0.000	−0.03	−3.45	5.82
I feel satisfied with my life	*0.004	−0.02	−2.56	2.77
I feel I have the opportunity to learn new things	0.051	−0.01	−1.23	2.12
I feel optimistic about my future	*0.048	−0.01	−1.14	2.24
I feel I can trust people	0.086	0.01	1.22	1.75
I feel happy with my life	*0.001	0.02	2.70	2.85

Appendix, Fig. 71 MWS results before and after visiting UK heritage places: mean score percentage change, standard deviation mean change, standard deviation percentage change, and P values.

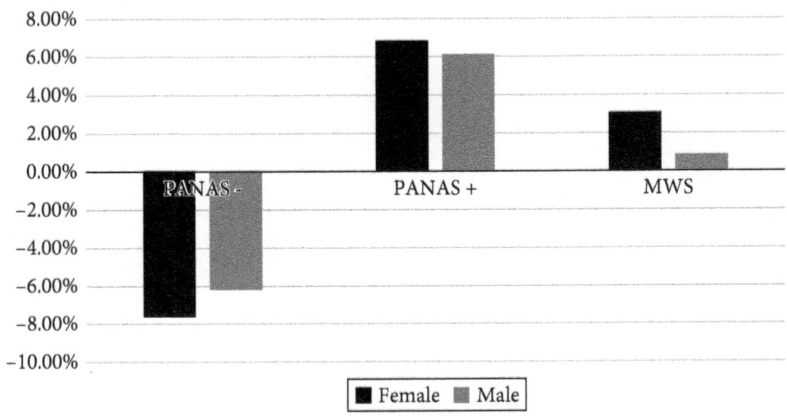

Appendix, Fig. 72 The impact of visitor gender at UK heritage places on mean average % PANAS positive and negative and MWS before and after scores.

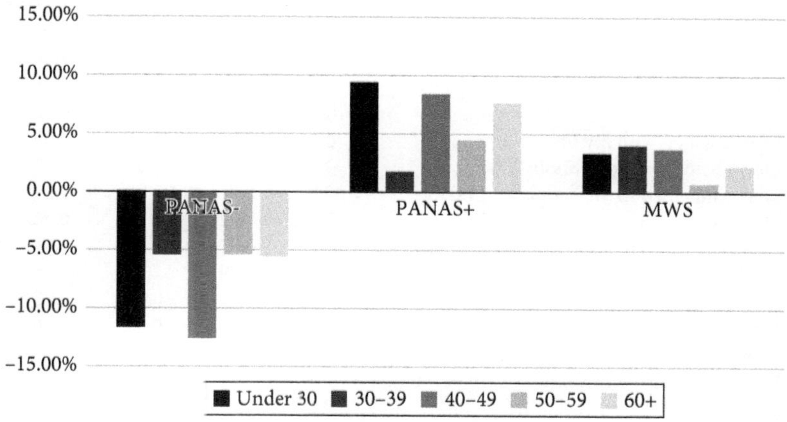

Appendix, Fig. 73 UK heritage place mean average % change before and after in PANAS+, PANAS–, and MWS for all age groups.

Measure	Survey	P Value	SD Change	SD % Change	Mean % Change
Distressed	PANAS –	*0.000	–0.09	–11.11	–8.43
Jittery	PANAS –	0.140	–0.06	–7.14	–3.61
Nervous	PANAS –	*0.016	–0.06	–6.98	0.27
Enthusiastic	PANAS +	0.658	–0.03	–3.61	–2.62
Guilty	PANAS –	0.179	–0.05	–6.85	–2.68
Irritable	PANAS –	*0.000	–0.03	–3.70	–6.35
Inspired	PANAS +	*0.014	–0.03	–3.45	2.00
Afraid	PANAS –	*0.029	–0.02	–2.82	–3.46
Scared	PANAS –	0.400	–0.01	–1.54	–1.26
Upset	PANAS –	*0.004	0	0.00	–3.67
Ashamed	PANAS –	0.304	0	0.00	0.83
Alert	PANAS +	0.606	–0.02	–2.74	0.48
Hostile	PANAS –	0.235	0.03	6.00	2.25
Determined	PANAS –	0.615	–0.01	–1.19	0.40
Active	PANAS –	0.954	–0.01	–1.08	0.41
Excited	PANAS –	0.443	0	0.00	0.63
Strong	PANAS –	0.137	0.01	1.16	1.02
Proud	PANAS –	0.307	0.01	1.18	0.96
Interested	PANAS +	0.811	0.01	1.56	0.03
Attentive	PANAS +	0.851	0.05	8.06	–0.06

Appendix, Fig. 74 PANAS results before and after visiting US heritage places: mean score percentage change, standard deviation percentage change, standard deviation mean change, and P value.

Measure	P Value	SD Change	SD % Change	Mean % Change
I feel interested in the world around me	0.759	−0.03	−4.55	0.06
I feel satisfied with my life	0.689	−0.03	−4.41	0.09
I feel I belong to a neighbourhood	0.560	−0.04	−4.00	−0.67
I feel connected to the people around me	*0.006	−0.03	−3.90	1.54
I feel a sense of accomplishment	0.235	−0.02	−2.74	−0.94
I feel connected to the local environment	0.211	−0.02	−2.50	0.98
I feel I have freedom of choice	*0.029	−0.02	−2.78	−1.16
I feel physically active	0.219	0	0.00	−0.67
I feel optimistic about my future	0.655	0	0.00	−0.16
I feel happy with my life	0.832	0	0.00	0.03
I feel I have the opportunity to learn new things	*0.004	0.01	1.59	−1.32
I feel people support each other	*0.000	0.02	2.50	3.36
I feel my life is worthwhile	0.082	0.02	3.13	−1.10
I feel I can trust people	*0.000	0.03	3.90	2.76
I feel I can overcome challenges	*0.047	0.03	4.55	−1.22

Appendix, Fig. 75 MWS results before and after US heritage places: mean score percentage change, mean standard deviation percentage change, standard deviation percentage change, and P values.

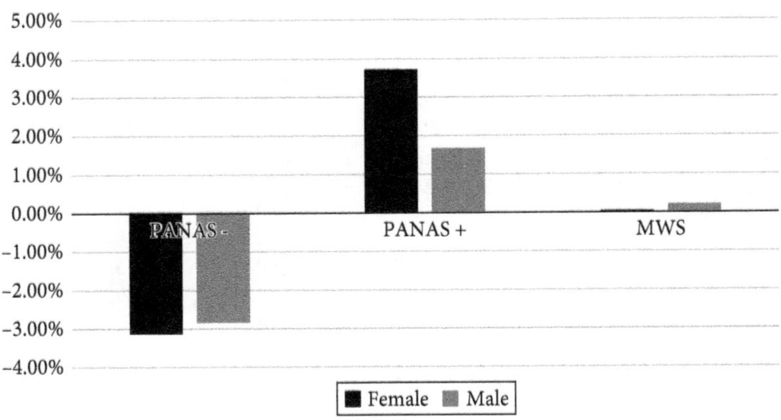

Appendix, Fig. 76 US heritage place mean average % change before and after in PANAS+, PANAS−, and MWS for female and male participants.

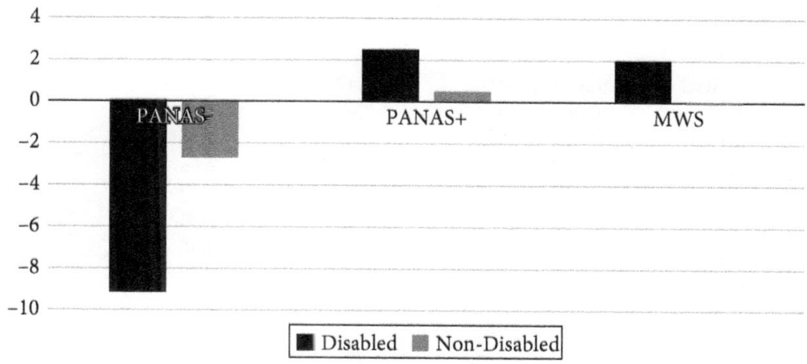

Appendix, Fig. 77 US heritage places mean average % change before and after in PANAS+, PANAS–, and MWS for disabled and non-disabled participants.

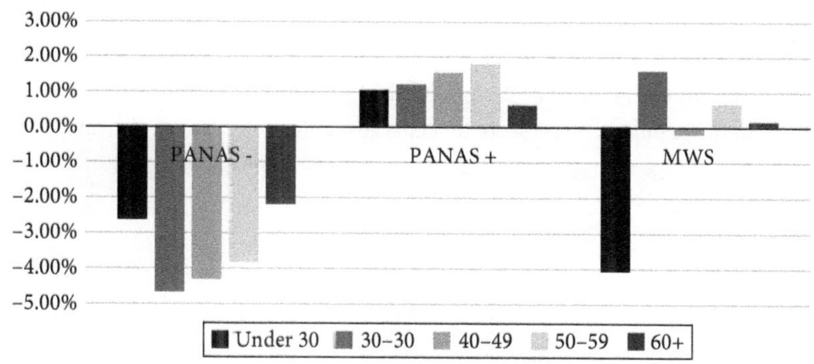

Appendix, Fig. 78 US heritage place mean average % change before and after in PANAS+, PANAS–, and MWS across all age groups.

Measure	P Value	SD Change	SD % Change	Mean % Change
Scared	*0.003	−0.03	−2.42	−7.71
Irritable	*0.005	0	0.00	−5.87
Upset	*0.006	−0.01	−0.83	−5.56
Jittery	0.235	0	0.00	−2.68
Ashamed	*0.033	0.04	3.42	−3.52
Afraid	0.282	0.06	5.00	−1.28
Nervous	0.548	0.07	5.93	−0.56
Guilty	0.305	0.08	6.84	−2.10
Distressed	0.113	0.08	6.72	−2.80
Hostile	0.225	0.1	8.33	−3.30
Excited	*0.000	−0.02	−1.87	11.02
Active	*0.000	−0.01	−0.83	9.35
Attentive	*0.000	0	0.00	8.81
Determined	*0.000	0.03	2.61	9.39
Interested	*0.000	0.04	3.70	15.70
Inspired	*0.000	0.06	5.88	11.05
Enthusiastic	*0.000	0.07	6.54	11.43
Strong	0.001	0.09	7.96	6.12
Alert	0.009	0.11	9.57	6.34
Proud	0.000	0.1	10.00	8.23

Appendix, Fig. 79 PANAS results before and after visiting Indian heritage places: mean score percentage change, standard deviation percentage change, standard deviation mean score change, and P value.

Measure	P Value	SD Change	SD Percentage Change	Mean % Change
I feel I can overcome challenges	*0.000	−0.04	−3.85	5.51
I feel optimistic about my future	*0.000	−0.02	−1.85	6.87
I feel satisfied with my life	*0.000	−0.01	−0.92	7.31
I feel connected to the people around me	*0.000	−0.02	−1.82	7.94
I feel interested in the world around me	*0.000	−0.01	−0.92	5.73
I feel I have freedom of choice	*0.032	0	0.00	2.55
I feel I have the opportunity to learn new things	*0.000	0.01	0.93	6.64
I feel I belong to a neighbourhood	*0.000	0.02	1.74	8.56
I feel connected to the local environment	*0.003	0.02	1.74	3.77
I feel my life is worthwhile	*0.000	0.02	1.74	6.43
I feel physically active	*0.000	0.03	2.70	5.38
I feel happy with my life	*0.000	0.05	4.59	4.51
I feel I can trust people	*0.010	0.07	5.34	3.94
I feel a sense of accomplishment	*0.000	0.07	6.73	6.58
I feel people support each other	*0.007	0.07	5.98	3.48

Appendix, Fig. 80 MWS results before and after visiting Indian heritage places: mean score percentage change, standard deviation percentage change, standard deviation change, and P value.

Bibliography

Abdallah, Saamah, Sorcha Manhony, Nic Marks, Juliet Michaelson, Charles Seaford, Laura Stoll, and Sam Thompson. 2011. *Measuring Our Progress: The Power of Wellbeing*. London: New Economics Foundation.

Abdallah, Saamah, Hanna Wheatley, and Annie Quick. 2017. *What Works Centre for Wellbeing: Drivers of Wellbeing Inequality*. Retrieved from https://whatworks wellbeing.org/product/drivers-of-wellbeing-inequality/ [accessed 28 November 2019].

Accenture. 2006. 'Capturing the Public Value of Heritage: Looking beyond the Numbers'. In *The Public Value of Heritage: The Proceedings of the London Conference 25–26 January 2006*, ed. Kate Clark. Swindon: English Heritage, 19–22.

Aked, Jody, Nic Marks, Cor Cordon, and Sam Thompson. 2008. *Five Ways to Wellbeing. A Report Presented to the Foresight Project on Communicating the Evidence Base for Improving People's Well-Being*. Centre for Wellbeing, The New Economics Foundation. Retrieved from https://neweconomics.org/uploads/files/five-ways-to-wellbeing-1.pdf [accessed 9 July 2020].

Alemdar, Melek, and Huseyin Anılan. 2020. 'The Development and Validation of the Emotional Literacy Skills Scale'. *International Journal of Contemporary Educational Research* 7(2): 258–270.

Allin, Paul, and David Hand. 2014. *The Wellbeing of Nations: Mean Motive and Measurement*. New York: Wiley.

Ander, Erica, Linda Thomson, Guy Noble, Anna Lanceley, Usha Menon, and Helen Chatterjee. 2013. 'Heritage, Health and Well-Being: Assessing the Impact of a Heritage Focused Intervention on Health and Well-Being'. *International Journal of Heritage Studies* 19(3): 229–242.

Annas, Julia. 1998. 'Virtue and Eudaimonism'. *Social Philosophy and Policy* 14(1): 37–55.

Asbury, Jonathan. 2016. *Imperial War Museum Duxford Guidebook*. London: Imperial War Museum.

Ascherson, Neal. 2000. 'Editorial'. *Journal of Public Archaeology* 1(1): 1–4.

Atalay, Sonya. 2007. 'Global Application of Indigenous Archaeology: Community Based Participatory Research in Turkey'. *Archaeologies* 249(3): 249–270.

Austin, Annie. 2018. 'Well-Being and Social Justice: In Defence of a Capabilities Approach'. In *The Politics of Wellbeing: Theory, Policy and Practice*, ed. Ian Bache and Karen Scott. London: Palgrave Macmillan, 49–70.

Bache, Ian, and Karen Scott. 2018. 'Wellbeing in Politics and Policy'. In *The Politics of Wellbeing: Theory, Policy and Practice*, ed. Ian Bache and Karen Scott. London: Palgrave Macmillan, 1–24.

Bahn, Paul. 1996. *Archaeology: A Very Short Introduction*. Oxford: Oxford University Press.

Belcher, Michael. 1991. *Exhibitions in Museums*. Leicester: Leicester University Press.

Belford, Paul. 2011. 'Archaeology, Community and Identity in a New English Town'. *Historic Environment* 2(1): 46–67.

Bender, Barbara. 1998. *Stonehenge: Making Space*. Oxford: Berg.

Biltmore Company. 2023. *Biltmore History*. Retrieved from https://www.biltmore.com/our-story/biltmore-history/ [accessed 20 March 2023].

Binford, Lewis. 1983. *In Pursuit of the Past. Decoding the Archaeological Record*. London: Thames and Hudson.

Blackburn, Simon. 2006. *Truth: A Guide for the Perplexed*. London: Penguin Books.

Bok, Sissela. 2010. *Exploring Happiness. From Aristotle to Brain Science*. New Haven: Yale University Press.

Bonacchi, Andrea, Franscesca Chiesi, Chloe Lau, Georgia Marunic, Donald Saklofske, Fabio Marra, and Guido Miccinesi. 2021. 'Rapid and Sound Assessment of Well-Being within a Multi-Dimensional Approach: The Well-Being Numerical Rating Scales (WB-NRSs)'. *PLoS One* 16(6): 1–21.

Bradley, Ben. 2015. *Wellbeing*. Cambridge: Polity Press.

Brave Heart, Maria Yellow Horse, and Lemyra DeBruyn. 1998. 'The American Indian Holocaust: Healing Historical Unresolved Grief'. *American Indian and Alaska Native Mental Health Research* 8(2): 60–82.

Brizi, Ambra, Anna Rabinovich, and Carenza Lewis. 2023. 'Psychological Outcomes of Local Heritage Engagement: Participation in community Archaeological Excavations Increases Well-Being, Self-Efficacy, and Perceived Community Support'. *Journal of Applied Social Psychology* 53(9): 850–861.

Bryman, Alan. 2012. *Social Research Methods*. Oxford: Oxford University Press.

Burtenshaw, Paul. 2014. 'Mind the Gap: Cultural and Economic Values in Archaeology'. *Journal of Public Archaeology* 13(1–3): 48–58.

Cabe. 2004. *The Value of Public Space: How High Quality Parks and Public Spaces Create Economic, Social and Environmental Value*. London: Commission for Architecture and the Built Environment.

Cabinet Office. 2013. *Guidance: Social Value Act: Information and Resources*. Retrieved from https://www.gov.uk/government/publications/social-value-act-information-and-resources/social-value-act-information-and-resources [accessed 2 January 2020].

Campion, Louise. 2022. 'The Holburne: Pathways to Wellbeing'. In *Creatively Minded at The Museum: Creative and Mental Health Activity in Museums*, ed. David Cutler. London: The Baring Foundation, 39–41.

Cane, James, Dennis O'Connor, and Susan Michie. 2012. Validation of the theoretical domains framework for use in behaviour change and implementation research. *Implementation Science* 7(37). https://doi.org/10.1186/1748-5908-7-37

Carman, John. 2012. Towards an International Comparative History of Archaeological Heritage Management. In *The Oxford Handbook of Public Archaeology*, ed. Robin Skeates, Carol McDavid, and John Carman. Oxford: Oxford University Press, 13–35.

Census. 2011. *Population Census Data Chandigarh*. Retrieved from https://www.census2011.co.in/census/city/22-chandigarh.html [accessed 20 March 2023].

Centre for Thriving Places. 2016. *Happy City Wellbeing Measurement: A Guide to Qualitative Data Collection*. Retrieved from https://www.centreforthrivingplaces.

org/wp-content/uploads/2016/10/Happy-City-Wellbeing-Measurement-A-Guide-to-Qualitative-Data-Collection.pdf [accessed 20 January 2022].

Chandigarh Government Museum and Art Gallery. 2020. *Chandigarh Government Museum and Art Gallery*. Retrieved from https://www.chdmuseum.gov.in/ [accessed 9 August 2022].

Chatsworth House Trust. 2023. *History of Chatsworth*. Retrieved from https://www.chatsworth.org/visit-chatsworth/chatsworth-estate/history-of-chatsworth [accessed 9 March 2023].

Chatterjee, Helen, and Paul Camic. 2015. 'The Health and Well-Being Potential of Museums and Art Galleries'. *Arts and Health* 7(3): 183–186.

Chi, Christina, Ruiying Cai, and Yongfen Li. 2017. 'Factors Influencing Residents Subjective Well-Being at World Heritage Sites'. *Tourism Management* 63: 209–222.

Christie, Ian, and Lindsay Nash. 1998. *The Good Life*. London: DEMOS.

Cleere, Henry. 1984. *Approaches to Archaeological Heritage*. Cambridge: Cambridge University Press.

Clifton, Briony. 2019. 'People and Making Places'. In *Historic Landscapes and Mental Well-Being*, ed. Tim Darvill, Kerry Barrass, Laura Drysdale, Vanessa Heaslip, and Yvette Staelens. Oxford: Archaeopress, 144–152.

Cohen, Deborah, Thomas McKenzie, Amber Sehgal, Stephanie Williamson, Daniela Golinelli, and Nicola Lurie. 2007. 'Contribution of Parks to Physical Activity'. *American Journal of Public Health* 97(3): 509–514.

Copeland, Tim. 2006. 'Presenting Archaeology to the Public: Constructing Insights On-Site'. In *Public Archaeology*, ed. Nick Merriman. London: Routledge, 132–144.

Council of Europe. 1950. *Convention for the Protection of Human Rights and Fundamental Freedoms*. Council of Europe Treaty Series 005. Paris: Council of Europe.

Council of Europe. 2005. *Council of Europe Framework Convention on the Value of Cultural Heritage for Society 2005*. Council of Europe Treaty Series – No. 199. Strasbourg: Council of Europe.

Crosby, Andrew. 2002. 'Archaeology and Vanua Development in Fiji'. *World Archaeology* 34(2): 363–378.

Cubitt, Geoffrey. 2007. *History and Memory*. Manchester: Manchester University Press.

Cutler, David. 2022. *Creatively Minded at the Museum: Creative and Mental Health Activity in Museums*. London: Baring Foundation. Retrieved from https://cdn.baringfoundation.org.uk/wp-content/uploads/BF_Creatively-minded-at-the-museum_WEB_mr.pdf [accessed 15 March 2023].

Dâmaso, Mafalda, Sophie Dowden, and Calum Smith. 2023a. *Compendium of sustainable culture-based solutions for well-being and health. CultureForHealth*. Culture Action Europe. Retrieved from https://static1.squarespace.com/static/56c99fffe32140b693299954/t/649594fd8814fd718d83c89e/1687524613088/C4H_Compendium.pdf [accessed 6 July 2022].

Dâmaso, Mafalda, Sophie Dowden, and Calum Smith. 2023b. *Cultural for Health and Wellbeing Compendium: Code for Practitioners*. Culture for Health.

Retrieved from https://www.cultureforhealth.eu/app/uploads/2023/06/C4H_Compendium_V4_LP.pdf [accessed 29 June 2023].

Darvill, Tim. 1995. 'Value Systems in Archaeology'. In *Managing Archaeology*, ed. Malcolm Cooper, Anthony Firth, John Carman and David Wheatley. London: Routledge, 40–50.

Darvill, Tim. 2019. 'Monuments for Life: Building Human Henge at Stonehenge and Avebury'. In *Historic Landscapes and Mental Well-Being*, ed. Tim Darvill, Kerry Barrass, Laura Drysdale, Vanessa Heaslip, and Yvette Staelens. Oxford: Archaeopress, 65–84.

Darvill, Tim, Kerry Barrass, Laura Drysdale, Vanessa Heaslip, and Yvette Staelens. 2019. *Historic Landscapes and Mental Well-Being*. Oxford: Archaeopress.

Darvill, Tim, Vanessa Heaslip, and Kerry Barrass. 2018. 'Heritage and Wellbeing'. In *Routledge Handbook for Wellbeing*, ed. Kathleen Galvin. London: Routledge, 112–123.

Dayson, Chris. 2013. *What Is Social Value and How Can It Be Commissioned*. Centre for Regional Economic and Social Research. Retrieved from https://shura.shu.ac.uk/11510/5/Dayson%20-%20Evaluating%20social%20innovations%20%28accepted%29.pdf [accessed 2 January 2020].

Department for Culture Media and Sport. 2016. *Taking Part 2015/16 Quarter 4 StatisticalRelease*. Retrieved from https://www.gov.uk/government/uploads/system/uploads/attachment_data/file/539312/Taking_Part_2015-16_Quarter_4_Report_-_FINAL.pdf [accessed 20 May 2017].

Department for Digital, Culture, Media and Sport. 2017. *Heritage Statement*. London: Department Digital, Culture, Media and Sport.

Department for Environment, Food, and Rural Affairs (DEFRA). 2009. *Sustainable Development Indicators in Your Pocket*. London: DEFRA.

Department for Environment, Transport, and the Regions (DETR). 2000. *Urban White Paper: Our Towns, Cities: The Future- Delivering an Urban Renaissance*. London: HMSO.

Department of Culture Media and Sport. 2006. *Better Places to Live: Government, Identity and the Public Value of Heritage. Summary of Responses*. London: Department of Culture, Media, and Sport.

Department of Health. 2009. *New Horizons. A Shared Vision for Mental Health*. London: Department of Health.

Department of Health. 2010a. *Our Health and Wellbeing*. London: Department of Health.

Department of Archaeology, Museums, and Heritage. 2023. Venkatappa Art Gallery Bengaluru. Retrieved from https://archaeology.karnataka.gov.in/page/Government+Museums/Venkatappa+Art+Gallery+Bengaluru/en [accessed 10 March 2023].

Department of Health. 2010b. *Healthy Lives, Healthy People: Our Strategy for Public Health in England*. London: Department of Health.

Department of Health. 2012. *Local Healthwatch: A Strong Voice for People—The Policy Explained*. London: Department of Health.

Desmarais, Sarah., Laura Bedford, & Helen Chatterjee. 2018. *Museums as Spaces for Wellbeing: A Second Report from the National Alliance for Museums, Health and Wellbeing*. Available at https://museumsandwellbeingalliance.wordpress.com

DeSilvey, Caitlin. 2017. *Curated Decay: Heritage Beyond Saving*. Minneapolis: University of Minnesota Press.

Diener, Ed. 2009a. *The Science of Well-Being: The Collected Works of Ed Diener*. New York: Springer Dordrecht.

Diener, Ed. 2009b. *Culture and Well-Being: The Collected Works of Ed Diener*. New York: Springer.

Diener, Ed, Marissa Diener, and Carol Diener. 1995. 'Factors Predicting the Subjective Well-Being of Nations'. *Journal of Personality and Social Psychology* 69(5): 851–864.

Dobat, Andres, Sultan Wood, Bo Jensen, Soren Schmidt, and Armin Dobat. 2019. '"I Now Look Forward to the Future, by Finding Things from our Past..." Exploring the Potential of Metal Detector Archaeology as a Source of Well-Being and Happiness for British Armed Forces Veterans with Mental Health Impairments'. *International Journal of Heritage Studies* 24(4): 370–386.

Dockery, Alfred. 2010. 'Culture and Wellbeing: The Case of Indigenous Australians'. *Social Indicators Research* 99(2): 35–332.

Dodge, Rachel, Annette Daly, Jan Huyton, and Lalage Sanders. 2012. 'The Challenge of Defining Wellbeing'. *International Journal of Wellbeing* 2(3): 222–235.

Donders, Yvonne. 2020. 'Cultural Heritage and Human Rights'. In *Oxford Handbook on International Cultural Heritage Law*, ed. Francesco Francioni and Ana Filipa Vrdoljak. Oxford: Oxford University Press, 335–357.

Doyal, Len, and Ian Gough. 1991. *A Theory of Human Need*. Basingstoke: MacMillian.

Drysdale, Laura. 2019. 'Walking with intent: Cultural therapy in ancient landscapes'. In *Historic Landscapes and Mental Well-Being*, ed. Tim Darvill, Kerry Barrass, Laura Drysdale, Vanessa Heaslip, and Yvette Staelens. Oxford: Archaeopress, 54–64.

Dulwich Picture Gallery. 2023. *About Our Collections*. Retrieved from https://www.dulwichpicturegallery.org.uk/about/ [accessed 9 March 2023].

Dunstable Town Council. 2018. *Priory Gardens Management Plan 2018–2024*. Retrieved from https://www.dunstable.gov.uk/wp/wp-content/uploads/2018/03/Priory-Gardens-Management-Plan-18-23.pdf [accessed 9 March 2023].

Eichstedt, Jennifer, and Stephen Small. 2002. *Representations of Slavery: Race and Ideology in Southern Plantation Museums*. Washington, DC: Smithsonian Books.

Emerson, Robert, Rachel Fretz, and Linda Shaw. 1995. *Writing Ethnographic Fieldnotes*. Chicago: University of Chicago Press.

English Heritage. 2000. *Power of Place: The Future of the Historic Environment*. London: English Heritage for the Historic Environment Steering Group.

English Heritage. 2004. *Heritage Counts*. Swindon: English Heritage.

English Heritage. 2006. *Heritage Counts: The State of England's Historic Environment 2006*. London: English Heritage.

English Heritage. 2014. *Heritage Counts*. Swindon: English Heritage.

European Commission. 2008. *European Pact for Mental Health and Well-Being*. Retrieved from https://health.ec.europa.eu/system/files/2016-11/euregha_support_statement_en_0.pdf [accessed 12 December 2019].

European Union. 2022. *EU Global Health Strategy: Better Health for All in a Changing World*. Luxembourg: European Union.

EuroQol Group. 1990. 'EuroQol: A New Facility for the Measurement of Health-Related Quality of Life'. *Health Policy* 16: 199–208.

Evans, Joanne, Ian Macrory, and Chris Randall. 2015. *Measuring National Wellbeing: Life in the UK, 2015*. Office for National Statistics. Retrieved from https://www.ons.gov.uk/peoplepopulationandcommunity/wellbeing/articles/measuringnationalwellbeing/2015-03-25 [accessed 12 December 2019].

Evans, Jules. 2018. 'The End of History and the Invention of Happiness'. In *The Politics of Wellbeing: Theory, Policy and Practice*, ed. Ian Bache and Karen Scott. London: Palgrave Macmillan, 25–48.

Everill, Paul, Richard Bennet, and Karen Burnell. 2020. 'Dig In: An Evaluation of the Role of Archaeological Fieldwork for the Improved Wellbeing of Military Veterans'. *Antiquity* 94(373): 212–227.

Everill, Paul, and Karen Burnell. 2022. *Archaeology, Heritage and Wellbeing: Authentic, Powerful and Therapeutic Engagement with the Past*. Abingdon: Routledge.

Fancourt, Daisy, Kamaldeep Bhui, Helen Chatterjee, Paul Crawford, Geoffrey Crossick, Tia DeNora, and Jane South. 2020. 'Social, Cultural and Community Engagement and Mental Health: Cross-Disciplinary, Co-produced Research Agenda'. *British Journal of Psychology Open* 7(1): E3. http://doi: 10.1192/bjo.2020.133.

Fancourt, Daisy, and Saoirse Finn. 2019. *Health Evidence Network Synthesis Report 67: What Is the Evidence on the Role of the Arts in Improving Health and Well-Being? A Scoping Review*. Copenhagen: World Health Organization.

Finch, Holmes. 2020. *Exploratory Factor Analysis*. London: Sage Publications.

Findlay, Jane, Kelly Robinson, and Alex Bowie. 2022. 'Dulwich Picture Gallery: Together through Art'. In *Creatively Minded at the Museum: Creative and Mental Health Activity in Museums*, ed. David Cutler. London: The Baring Foundation, 27–29.

Finnegan, Alan. 2016. 'The biopsychosocial benefits and shortfalls for armed forces veterans engaged in archaeological activities'. *Nurse Education Today* 47: 15–22.

Finneran, Niall, and Christina Welch. 2022. Having a Wander through Whitechapel: Towards a Methodological Framework for a Therapeutic Urban Psychography. In *Archaeology, Heritage and Wellbeing: Authentic, Powerful and Therapeutic Engagement with the Past*, ed Paul Everill, and Karen Burnell. Abingdon: Routledge, 216–273.

Fluck, Helen. 2015. *Health and Wellbeing and the Historic Environment, Historic England Horizon Scan 2015–14*. Swindon: Historic England.

Forgeard, Marie, Eranda Jayawickreme, Margaret Kern and Martin Seligman. 2011. 'Doing the Right Thing: Measuring Wellbeing for Public Policy'. *International Journal of Wellbeing*, 1, 79–106.

Fredericksen, Clayton. 2002. 'Caring for History: Tiwi and Archaeological Narratives of Fort Dundas/Punata, Melville Island, Australia'. *World Archaeology* 34(2): 288–302.

Friedman, Milton. 1962. *Capitalism and Freedom*. Chicago: University of Chicago Press.

Fujiwara, Daniel, Thomas Cornwall, and Paul Dolan. 2014. *Heritage and Wellbeing* (unpublished report). Swindon: English Heritage.

Fujiwara, Daniel, and George MacKerron. 2015. *Cultural Activities, Artforms and Wellbeing, Arts Council England*. Retrieved from https://www.artscouncil.org.

uk/sites/default/files/download-file/Cultural_activities_artforms_and_wellbeing.pdf [accessed 18 October 2017].

Galeotti, Glenda. 2016. 'Elements for Impact Assessment of Cultural Heritage and Community Wellbeing: A Qualitative Study on Casentino's Eco Museum'. *Journal of the Section of Cultural Heritage* 14: 913–942.

Gallou, Eirini. 2022. 'Heritage and Pathways to Wellbeing'. *Studies of a Value of Cultural Heritage* 14: 913–943.

Gallou, Eirini, and Kalliopo Fouseki. 2019. 'Applying Social Impact Assessment (SIA) Principles in Assessing the Contribution of Cultural Heritage to Social Sustainability in Rural Landscapes'. *Journal of Cultural Heritage Management and Sustainable Development* 9(3): 352–375.

Garcia, Danielle, and Andrea Winn. 2017. *Inspiring Futures: Volunteering for Wellbeing: Final Report 2013–2016—Social Return on Investment*. Retrieved from http://volunteeringforwellbeing.org.uk/wp- content/uploads/2017/03/IF_VOLUNTEERING_FOR_WELLBEING_REPORT_2013-16_SROI_IWM.pdf [accessed 29 October 2017].

Gee, Graham, Pat Dudgeon, Clinton Schultz, Amanda Hart, and Kerrie Kelly. 2014. 'Social and Emotional Wellbeing and Mental Health: An Aboriginal Perspective'. In *Working Together: Aboriginal and Torres Strait Islander Mental Health and Wellbeing Principle and Practice*, ed. Pat Dudgeon, Helen Milroy, and Roz Walker. Canberra: Australian Government Department of the Prime Minister and Cabinet, 55–68.

German Federal Government. 2018. *Government Report on Wellbeing in Germany*. Retrieved from https://www.gut-leben-in-deutschland.de/SharedDocs/Downloads/EN/LB/Government-Report-on-Wellbeing-in-Germany.pdf [accessed 8 December 2019].

Gesler, Wilbert. 2005. 'Therapeutic Landscapes: An Evolving Theme'. *Health and Place* 11(4): 295–297.

Giliberto, Francesca. 2021. UNESCO. 'Heritage, Mental Health and Wellbeing'. *Heritage and Our Sustainable Futures* 6. Retrieved from https://unesco.org.uk/wp-content/uploads/2021/11/Mental-Health-and-Wellbeing-Report.pdf [accessed 6 February 2023].

Goldberg, David, and Valerie Hillier. 1979. 'A scaled version of the General Health Questionnaire'. *Psychol Med.* 9(1): 139–45.

Government of Karnataka. 2023. Bengaluru Urban District. Retrieved from https://bengaluruurban.nic.in/en/history/ [accessed 20 March 2023].

Graham, Carol. 2011. *The Pursuit of Happiness: An Economy of Wellbeing*. Washington, DC: The Brookings Institute Press.

Greer, Shelley, Rodney Harrison, and Susan McIntyre-Tamwoy. 2002. 'Community Based Archaeology in Australia'. *World Archaeology* 34(2): 265–287.

Griffin, James. 1988. *Well-Being: Its Meaning, Measurement and Moral Importance*. Wotton-Under-Edge: Clarendon Press.

Haidt, Jonathan. 2007. *The Happiness Hypothesis: Putting Ancient Wisdom to the Test of Modern Science*. London: Arrow.

Halpern, David. 2015. *Inside the Nudge Unit: How Small Changes Can Make a Difference*. London: WH Allen.

Halton NHS Clinical Commissioning Group. 2017. *A Cultural Manifesto for WellBeing*. Retrieved from http://www.atmag.org/wp-content/uploads/2018/05/A-Cultural-Manifesto-for-Wellbeing.pdf [accessed 28 November 2019].

Hamilakis, Yannis, and Theou Efthimis. 2013. *Enacted Multi-Temporality: The Archaeological Sites as a Shared Performative Space. Reclaiming Archaeology: Beyond the Tropes of Modernity, in Alfredo Gonzalez-Ruibal*. London: Routledge.

Hammett, Amy, and Emma Gascoigne. 2022. 'The Salisbury Museum: Well-City Salisbury'. In *Creatively Minded at the Museum: Creative and Mental Health Activity in Museums*, ed. David Cutler. London: The Baring Foundation, 49–51.

Happy Museum Project. 2011. *About*. Retrieved from http:/happymuseumproject.org/about/ [accessed 12 December 2019].

Harrison, Rodney. 2013. *Heritage Critical Approaches*. London: Routledge.

Harrison, Rodney, Caitlin DeSilvey, Cornelius Holtorf, Sharon Macdonald, Nadia Bartolini, Esther Breithoff, Harald Fredheim, Antony Lyons, Sarah May, Jennie Morgan, and Sefryn Penrose. 2020. *Comparative Approaches to Natural and Cultural Heritage Practices*. London: University College London Press.

Heaslip, Vanessa. 2019. 'Human Henge: The Impact of Neolithic Healing Landscapes on Mental Health and Well-Being'. In *Historic Landscapes and Mental Well-Being*, ed. Tim Darvill, Kerry Barrass, Laura Drysdale, Vanessa Heaslip, and Yvette Staelens. Oxford: Archaeopress, 123–134.

Hems, Alison. 2006. 'Introduction: Beyond the Graveyard—Extending Audiences, Enhancing Understanding'. In *Heritage Interpretation*, ed. Alison Hems and Marion Blockley. London: Routledge, 1–8.

Hendry, Joy. 1999. *An Introduction to Social Anthropology: Other People's World*. London: Macmillan Press Ltd.

Henson, Don. 2004. 'The Education Framework in the United Kingdom'. In *Education and the Historic Environment*, ed. Don Henson, Peter Stone, and Mike Corbishley. London: Routledge, 13–22.

Heritage Alliance. 2020. *Heritage, Health and Wellbeing: A Heritage Alliance Report*. London: Heritage Alliance.

Heritage Lottery Fund. 2015. *All Our Stories Evaluation*. Retrieved from https://www.heritagefund.org.uk/about/insight/evaluation/all-our-stories-evaluation [accessed 12 September 2019].

Heritage Lottery Fund. 2018. *Wellbeing Guidance*. Retrieved from https://www.heritagefund.org.uk/publications/wellbeing-guidance [accessed 12 September 2019].

Hewison, Robert. 1987. *The Heritage Industry: Britain in a Climate of Decline*. London: Routledge.

Hewison, Robert, and John Holden. 2006. 'Public Value as a Framework for Analysing the Value of Heritage: The Ideas'. In *The Public Value of Heritage: The Proceedings of the London Conference 25–26th January 2006*, ed. Kate Clarke. Swindon: English Heritage, 14–18 and 19–22.

Hills, Catherine. 2001. 'The Dissemination of Information'. In *Archaeological Resource Management in the UK: An Introduction*, ed. Jon Hunter and Ian Ralston. Stroud: Sutton Publishing, 215–224.

Historic England. 2017. *Heritage and Society 2017*. Retrieved from https://historicengland.org.uk/content/heritage-counts/pub/2019/heritage-and-society-2019/ [accessed 30 October 2017].

Historic England. 2022. *Wellbeing and Heritage Strategy*. Retrieved from https://historicengland.org.uk/content/docs/about/strategy-wellbeing-heritage-2022-25/ [accessed 23 February 2023].

Hodder, Ian. 1986. *Reading the Past: Current Approaches to Interpretation in Archaeology*. Cambridge: Cambridge University Press.

Hodder, Ian. 2000. *Towards Reflexive Method in Archaeology: The Example at Çatalhöyük*. Cambridge: McDonald Institute for Archaeological Research, University of Cambridge.

Hodder, Ian. 2010. 'Cultural Heritage Rights: From Ownership and Descendent Justice and Wellbeing'. *Anthropology Quarterly* 83(4): 861–882.

Holtorf, Cornelius. 2005. *From Stonehenge to Las Vegas: Archaeology as Popular Culture*. Oxford: Altamira Press.

Holtorf, Cornelius. 2006. *Archaeology is a Brand! The Meaning of Archaeology in Contemporary Popular Culture*. Oxford: Archaeopress.

Holtorf, Cornelius. 2018. 'Embracing Change: How Cultural Resilience Is Increased through Cultural Heritage'. *World Archaeology* 50(4): 639–650.

Hooper-Greenhill, Eilean. 2004. *Inspiration, Identity, Learning: The Value of Museum: The Evaluation of the Impact of/DfES Strategic Commissioning 2003–2004*. Leicester: National/Regional Museums Education Partnerships. Retrieved from file:///C:/Users/sayerfa/Downloads/Inspiration%20Identity%20Learning%20The%20value%20of%20museums%20(1).pdf [accessed 5 September 2013].

Hooper-Greenhill, Eilean. 2007. *Museums and Education: Purpose, Pedagogy, Performance*. London: Routledge.

Howard, Peter. 2003. *Heritage, Management, Interpretation, Identity*. London: Continuum.

Human Dignity and Humiliation Studies. 2008. *Dignity Index: A Developing Proposal*. Retrieved from https://emotionalcompetency.com/hdi/index.htm [accessed 2 January 2021].

Human Henge. 2018 *Historic Landscape and Mental Health at Stonehenge and Avebury*. Retrieved from http://humanhenge.org [accessed 2 January 2020].

Imperial War Museum. 2023. *What We Spend and How We Spend It*. Retrieved from https://www.iwm.org.uk/corporate/foi/transparency [accessed 20 March 2023].

Isherwood, Rob. 2009. *Community Archaeology: A Study of the Conceptual, Political and Practical Issues Surrounding Community Archaeology in the United Kingdom Today*. Unpublished Doctoral Thesis. The University of Manchester.

Jeppson, Patrice. 2008. 'Doing Our Homework: Rethinking the Goals and Responsibilities of Archaeology Outreach to Schools'. In *Changing the World with Archaeology: Activist Archaeology*, ed. Jay Stottman. Greenville: University of Florida Press, 1–58.

Johnson, Matthew. 1999. *Archaeological Theory: An Introduction*. London: Blackwell.

Johnston, Helen. 2019. 'Messing about on the river: Volunteering and well-being on the Thames foreshore'. In *Historic Landscapes and Mental Well-Being*, ed. Tim Darvill, Kerry Barrass, Laura Drysdale, Vanessa Heaslip, and Yvette Staelens. Oxford: Archaeopress, 179–188.

Jones, Andrew. 1995. Integrating School Visits, Tourists and the Community at the Archaeological Resource Centre, York, UK'. In *Museum, Media, Message*, ed. Eilean Hopper-Greenhill. Leicester: Leicester University Press, 156–164.

Jones, Andrew, 2004. 'Using Objects: The York Archaeological Trust Approach'. In *Education and the Historic Environment*, ed. Don Henson, Peter Stone, and Michael Corbishley. London: Routledge, 173–184.

Jones, Sian. 2004. *Early Medieval Sculpture and the Production of Meaning, Value and Place: The Case of Hilton Cadboll. Historic Scotland*. Edinburgh: Historic Scotland.

Jones, Sian. 2004. *Research Report: Early Medieval Sculpture and the Productions of Meaning, Values and Place: The Case of Hilton of Cadboll*. Edinburgh: Historic Scotland.

Jordon, Bill. 2008. *Welfare and Well-Being: Social Value in Public Policy*. Bristol: Policy Press.

Kahneman, Daniel, Ed Diener, and Norbert Schwarz. 1999. *Wellbeing: Foundations of Hedonic Psychology*. London: Russell Sage Foundation.

Keeney, Sinead, Felicity Hasson, and Hugh McKenna. 2010. *The Delphi Technique in Nursing and Health Research*. London: Wiley.

Kiddey, Rachel. 2017. *Homeless Heritage: Collaborative Social Archaeology as Therapeutic Practice*. London: Routledge.

Kiddey, Rachel, and John Schofield. 2011. 'Embrace the Margins: Adventures in Archaeology and Homelessness'. *Public Archaeology* 10(1): 4–22.

Kindleysides, Michelle, and Emma Biglands. 2015. '"Thinking outside the Box, and Making It Too": Piloting an Occupational Therapy Group at an Open-Air Museum'. *Arts & Health* 7(3): 271–278.

Knight, Anneyce, Vincent La Placa, and Allan McNaught. 2014. *Wellbeing: Policy and Practice*. Banbury: Scion Publishing.

Kroenke, Kurt, Robert Spitzer, Janet Williams, and Bernd Lowe. 2010. 'The Patient Health Questionnaire Somatic, Anxiety, and Depression Symptom Scales: A Systematic Review'. *General Hospital Psychiatry* 32(4): 345–359.

Kroenke, Kurt, and Janet Williams. 2006. 'A Brief Measure for Assessing Generalized Anxiety Disorder: The GAD-7'. *Journal of the American Medical Association International Medicine* 166(10):1092–1097.

Laerd Statistics. 2018. *Mann-Whitney U Test Using SPSS Statistics*. Retrieved from https://statistics.laerd.com/spss-tutorials/mann-whitney-u-test-using-spss-statistics.php [accessed 2 January 2020].

Lakey, Jane, Neil Smith, Anni Oskala, and Sally McManus. 2017. *Culture, Sport and Wellbeing: Findings from the Understanding Society Youth Survey*. London: NatCen Social Research. Retrieved from https://www.artscouncil.org.uk/sites/default/files/download-file/Culture%20sport%20and%20wellbeing%20-%20youth.pdf [accessed 28 November 2019].

Lammy, David. 2006. 'Community, Identity and Heritage'. In *Capturing the Public Values of Heritage: The Proceedings of the London Conference 25–26 January 2006*, ed. Kate Clark. Swindon: English Heritage, 65–69.

LaRoche, Cheryl. 2005. 'Heritage, Archaeology and African American History'. *SAA Archaeological Record* 5(2): 34–37.

Lassibille, Gerard, and Ma Lucia Navarro Gómez. 2014. 'Human Capital'. In *Encyclopedia of Quality of Life and Well-Being Research*, ed. Alex Michalos. Dordrecht: Springer, 2999–3002.

Layton, Robert. 1989. *Who Needs the Past? Indigenous Values and Archaeology*. London: Unwin Hyman.

Liddle, Ian, and Greg Carter. 2015. 'Emotional and Psychological Well-Being in Children: The Development and Validation of the Stirling Children's Well-Being Scale'. *Educational Psychology in Practice* 31(2): 174–185.

Lipe, William. 1984. 'Value and Meaning in Cultural Resources'. In *Approaches to the Archaeological Heritage: A Comparative Study of World Cultural Resource Management Systems*, ed. Henry Cleere. Cambridge: Cambridge University Press, 1–11.

Lipe, William. 2007. 'Value and Meaning in Cultural Resources'. In *Cultural Heritage: Critical Concepts in Media and Cultural Studies*, Vol. I, ed. Laura-Jane Smith. London: Routledge, 286–306.

Lowenthal, David. 1985. *The Past is a Foreign County*. Cambridge: Cambridge University Press.

Lowenthal, David. 1998. *The Heritage Crusade and the Spoils of History*. Cambridge: Cambridge University Press.

Lucas, Richard. 2018. 'Reevaluating the Strengths and Weaknesses of Self-Report Measures of Subjective Well-Being'. In *Handbook of Well-Being*, ed. Ed Diener, Shigehiro Oishi, and Louis Tay. Salt Lake City: DEF Publishers, 45–55.

Lyubomirsky, Sonja, Kennon Sheldon, and David Schkade. 2005. 'Pursuing Happiness: The Architecture of Sustainable Change'. *Review of General Psychology* 9(2): 111–131.

Mackenzie, Robert, and Peter Stone. 1990. 'Introduction: A Concept of Excluded Past'. In *The Excluded Past*, ed. Peter Stone and Robert Mackenzie. London: Unwin Hyman Ltd, 1–14.

Maslow, Abraham. 1943. 'The Theory of Human Motivation'. *Psychological Review* 50: 370–396.

Mathews, Gordon, and Carolina Izquierdo. 2009. *Pursuits of Happiness: Well-Being in Anthropological Perspective* New York: Berghahn Books.

McClanahon, Angela. 2006. 'Histories, Identity, and Ownership: An Ethnographic Case Study in Archaeological heritage Management in the Orkney Islands'. In *Ethnographies of Archaeological Practice: Cultural Encounters, Material Transformations*, ed. Mathew Edgeworth. Lanham: Altamira Press, 126–136.

McGimsey, Charles. 1972. *Public Archaeology*. New York: McGraw Hill.

McNaught, Allan, and Anneyce Knight. 2011. *Understanding Wellbeing: An Introduction for Students and Practitioners of Health and Social Care*. Oxford: Lantern Publishing.

Mehmet, Nevin. 2011. 'Ethics and Wellbeing'. In *Understanding Wellbeing: An Introduction for Students and Practitioners of Health and Social Care*, ed. Allan Knight and Anneyce McNaught. Banbury: Lantern Publishing, 37–49.

Merriman, Nick. 1991. *Beyond the Glass Case: The Past, the Heritage and the Public in Britain*. Leicester: Leicester University Press.

Meskell, Lynn. 2011. *The Nature of Heritage: The New South Africa*. London: Wiley Blackwell.

Michaelson, Juliet, Saamah Abdallah, Nicola Steuer, Sam Thompson, Nic Marks, Jody Aked, Corrina Cordon, and Ruth Potts. 2009. *National Account of*

Well-Being: Bringing Real Wealth onto the Balance Sheet. London: New Economics Foundation.

Michie, Susan, Maartje Van Stralen, and Robert West. 2011. 'The Behaviour Change Wheel: A New Method for Characterising and Designing Behaviour Change Interventions'. *Implementation Science* 6(42): 1–11.

Michie, Susan, Lou Atkins. and Robert West. 2014b. *The Behaviour Change Wheel: A Guide to Designing Interventions*. Sutton: Silverback Publishing.

Michie, Susan, Robert West, Rona Campbell, Jamie Brown, and Heather Gainforth. 2014a. *ABC of Behaviour Change Theories*. Sutton: Silverback Publishing.

Ministry for Culture and Heritage. 2002. *Cultural Wellbeing*. New Zealand Government. Retrieved from https://mch.govt.nz/what-we-do/our-projects/cultural-well-being [accessed 28 June 2023].

Monkton, Linda. 2022. 'Wellbeing and the Historic Environment: A Strategic Approach'. In *Archaeology, Heritage and Wellbeing: Authentic, Powerful and Therapeutic Engagement with the Past*, ed. Paul Everill and Karen Burnell. Abingdon: Routledge, 239–260.

Munjeri, Dawson. 2006. 'Tangible and Intangible Heritage'. In *Cultural Heritage: Critical Concepts in Media and Cultural Studies*, ed. Laura-Jane Smith. London: Routledge, 323–330.

National Trust. 2018. *Annual Report 2017–2018*. Retrieved from https://gat04-live-1517c8a4486c41609369c68f30c8-aa81074.divio-media.org/filer_public/ec/35/ec35c337-20b6-43f5-9204-d84dec7163a8/rvr-08-full.pdf [accessed 28 November 2019].

Nemours Foundation. 2023. *Nemours Estate*. Retrieved from https://nemoursestate.org/dupont.html [accessed 20 March 2023].

New Zealand Government. 2019. *Local Government (Community Well-being) Amendment Bill. Government Bill 48–2*. Retrieved from http://www.legislation.govt.nz/bill/government/2018/0048/latest/whole.html#LMS30972 [accessed 10 December 2019].

Newman, William. 1995. *Social Research Methods: Qualitative and Quantitative Approach*. Boston: Allyn and Bacon.

Nimenko, Wasyl, and Roger Simpson. 2014. 'Rear Operations Group Medicine: A Pilot Study of Psychological Decompression in a Rear Operations Group during Operation HERRICK 14'. *Journal of Royal Army Medical Corps* 160: 295–297.

Nissinaho, Aino, and Tuija-Liisa Soininen. 2014. 'Adopt a Monument: Social Meaning from Community Archaeology'. In *Public Participation in Archaeology*, ed. Suzie Thomas and Joanna Lea. Woodbridge: Boydell Press, 175–182.

Nussbaum, Martha. 2005. 'Wellbeing, Contracts and Capabilities'. In *Rethinking Wellbeing*, ed. Leonore Manderson. Perth, Australia: API Network, 27–44.

Nussbaum, Martha, and Amartya Sen. 1993. *The Quality of Life*. New York: Oxford University Press.

Office of National Statistics. 2011. *Census UK 2011*. Retrieved from https://www.ons.gov.uk/census/2011census [accessed 9 December 2021].

Office of National Statistics. 2013. *Measuring National Well-being, Personal Wellbeing in the UK 2012/13*. London: Office of National Statistics. Retrieved from https://webarchive.nationalarchives.gov.uk/ukgwa/20160105231849/http://www.ons.gov.uk/ons/rel/wellbeing/measuring-national-well-being/personal-well-being-in-the-uk--2012-13/index.html [accessed 9 December 2019].

Office of National Statistics. 2018. *Measure of National Wellbeing; Quality of Life in the UK 2018*. London: Office of National Statistics. Retrieved from https://www.ons.gov.uk/peoplepopulationandcommunity/wellbeing/articles/measuringnationalwellbeing/qualityoflifeintheuk2018 [accessed 9 December 2019].

Office of National Statistics. 2021. *Census UK, 2021*. Retrieved from https://www.ons.gov.uk/census [accessed 10 January 2023].

Organization for Economic Co-operation and Development. 2001. *The Well-Being of Nations: The Role of Human and Social Capital*. Paris: OECD. Retrieved from https://www.oecd-ilibrary.org/docserver/9789264189515-en.pdf?expires=1680000371&id=id&accname=guest&checksum=F66365CC27323277027A65CD273B4A0C [accessed 12 December 2022].

Organization for Economic Co-operation and Development. 2017. *How's Life? Measuring Wellbeing*. Paris. OECD. Retrieved from https://doi.org/10.1787/how_life-2017-en [accessed 9 December 2019].

Paddon, Hannah, Linda Thomson, Usha Menon, Anne Lanceley, and Helen Chatterjee. 2014. 'Mixed Methods Evaluation of Well-Being Benefits Derived from a Heritage-in-Health Intervention with Hospital Patients'. *Arts & Health* 6(1): 24–58.

Pearson, Vikki. 2001. *Teaching the Past: A Practical Guide for Archaeologists*. York: Council for British Archaeology.

Pennington, Andy, Rebecca Jones, Anne-Marie Bagnall, Jane South, and Rhiannon Corcoran. 2018. *Heritage and Wellbeing: The Impact of Historic Places and Assets on Community and Wellbeing—A Scoping Review*. London: What Works Centre for Wellbeing.

Pfizer. (n.d.). *Instruction manual: Instructions for Patient Health Questionnaire (PHQ) and GAD-7 measures*. Retrieved from http://www.phqscreeners.com/instructions/instructions.pdf

Pfizer. 2010. *Patient Health Questionnaire (PHQ-9)*.

Pfizer. 2017. *Welcome to the Patient Health Questionnaire (PHQ) Screeners*. Retrieved from https://www.phqscreeners.com/select-screener/ [accessed 7 January 2020].

Phillips, Laura. 2008. 'Reminiscence: Recent Work at the British Museum'. In *Touch in Museums: Policy and Practice in Object Handling*, ed. Helen Chatterjee. Oxford: Berg, 199–204.

Pirson, Michael. 2017. *Humanistic Management: Protecting Dignity and Promoting Well-Being*. Cambridge: Cambridge University Press.

Power, Andrew, and Karen Smyth. 2016. 'Heritage, Health and Place: The Legacies of Local Community-Based Heritage Conservation on Social Wellbeing'. *Health & Place* 39: 160–167.

Prentice, Richard. 2001. 'Experiential Cultural Tourism: Museums and the Marketing of New Romanticism on Evoked Authenticity'. *Museum Management and Curatorship* 19(1): 5–26.

Prentice, Richard, Stephen Witt, and Claire Hamer. 1998. 'Tourism as experience: The case of heritage parks'. *Annals of Tourism Research*, Volume 25 (1) Pages 1–24, https://doi.org/10.1016/S0160-7383(98)00084-X

Price, Max, and Sophie Keynes. 2020. *Heritage, Health and Wellbeing*. Heritage Alliance. Retrieved from https://www.theheritagealliance.org.uk/wp-content/uploads/2020/10/Heritage-Alliance-AnnualReport_2020_Online.pdf [accessed 22 January 2023].

Putnam, Hilary. 2006. 'Objectivity and the Science-Ethics Distinction'. In *The Quality of Life*, ed. Martha Nussbaum and Amartya Sen. Oxford: Clarendon Press, 143–157.

Rabbas, Oyvind, Eyjolfur Emilson, Hallvard Fossheim, and Miira Tuominen. 2015. 'Introduction'. In *The Quest for the Good Life: Ancient Philosophers on Happiness*, ed. Oyvind Rabbas, Eyjolfur Emilson, Hallvard Fossheim, and Miira Tuominen. Oxford: Oxford University Press.

Rasbery, Sonia, and Jocelyn Goddard. 2011. *Mental Health and Heritage Working in Partnership: 4 Linked Projects in Surrey, MLA Renaissance Southwest.* Retrieved from https://abcofworkingwithschools.files.wordpress.com/2011/06/mental-health-heritage-working-in-partnership.pdf [accessed 2 January 2020].

Rathouse, William. 2019. 'Archaeology and Mental Health: War Memorial Survey in Ceredigion'. In *Historic Landscapes and Mental Well-Being*, ed. Tim Darvill, Kerry Barrass, Laura Drysdale, Vanessa Heaslip, and Yvette Staelens. Oxford: Archaeopress, 242–253.

Reeve, John and Vicky Woollard. 2007. 'Influences on Museum Practice'. In *The Responsive Museum: Working with Audiences in the 21st Century*, ed. Caroline Lang, John Reeve, and Vicky Woollard. London: Routledge, 211–222.

Reily, Sarah, Claire Nolon, and Linda Monckton. 2018. *Wellbeing and the Historic Environment*. Swindon: Historic England.

Renfrew, Colin. 2006. *Figuring It Out: What Are We? Where Do We Come From.* London: Thames and Hudson.

Ripanti, Francesco. 2022a. *Unforgettable Encounters: Understanding Participation in Italian Community Archaeology*. Oxford: Achaeopress.

Ripanti, Francesco. 2022b. LOGGIA. Retrieved from https://www.loggia-project.eu/ [accessed 10 December 2023].

Rizzo, Ilde, and Ruth Towse. 2002. *The Economics of Heritage*. Cheltenham: Edward Elgar Publishing.

Rose, David. 2009. *Aristotle. The Nicomachean Ethics*. Oxford: Oxford University Press.

Rosemberg, Cristina, Richard Naylor, Ulrike Chouguley, Lucy Mantella, and Kate Oakley. 2011. *Final Report: Assessment of the Social Impact of Volunteering in HLF-funded projects: Year.3*. Retrieved from https://www.heritagefund.org.uk/sites/default/files/media/research/social_impact_volunteering_2011.pdf [accessed 4 January 2020].

Ryan, Mary. 2019. *Pathways to Wellbeing Evaluation 2016–19*. Retrieved from https://www.holburne.org/wp-content/uploads/2022/07/Museums-%E2%80%93-Changing-lives-through-art-2016-19-1.pdf [accessed 15 March 2023].

Ryan, Mary. 2022. *Pathways to Wellbeing Evaluation 2019–22*. Retrieved from https://www.holburne.org/wp-content/uploads/2022/08/Museums-of-the-future.-Caring-for-communities-as-well-as-collections-2019–22-.pdf [accessed 15 March 2023].

Ryan, Richard, and Edward Deci. 2001. 'On Happiness and Human Potentials: A Review of Research on Hedonic and Eudaimonic Well-Being'. *Annual Review of Psychology* 52: 141–166.

Ryff, Carol. 1989. 'Happiness Is Everything, or Is It? Exploration of the Meaning of Psychological Well-Being'. *Journal of Personality and Social Psychology* 57: 1069–1081.

Sagger, Harman, Jack Philips, and Mohammed Haque. 2021. *Valuing Culture and Heritage Capital: A Framework towards Informing Decision Making*. London: DCMS. Retrieved from https://www.gov.uk/government/publications/valuing-culture-and-heritage-capital-a-framework-towards-decision-making/valuing-culture-and-heritage-capital-a-framework-towards-informing-decision-making [accessed 1 March 2023].

Sayer, Faye. 2014. 'Politics and the Development of Community Archaeology'. *Historic Environment: Policy and Practice* 7(1): 55–73.

Sayer, Faye. 2015. 'Can Digging Make You Happy? Archaeological Excavations, Happiness and Heritage'. *Arts and Health* 7(3): 247–260.

Sayer, Faye. 2018. 'Understanding Well-Being: A Mechanism for Measuring the Impact of Heritage Practice on Well-Being'. In *The Oxford Handbook of Public Heritage Theory and Practice*, ed. Angela Labrador and Neil Silberman. Oxford: Oxford University Press.

Sayer, Faye. 2019. *Public History: A Practical Guide*. London: Bloomsbury.

Sayer, Faye. 2022. *Public History*. Bloomsbury History: Theories and Methods. Retrieved from https://www.bloomsburyhistorytheorymethod.com/article?docid=b-9876543210999&tocid=b-9876543210999-004 [accessed 3 March 2023].

Sayer, Faye, and Amy Luck 2022. *Annual Evaluation Report Well-City Salisbury*. Unpublished.

Sayer, Faye, and Duncan Sayer. 2016. 'Bones without Barriers: The Social Impact of Digging the Dead'. In *Archaeologists and the Dead: Mortuary Archaeology and Contemporary Society*, ed. Mel Giles and Howard Williams. Oxford: Oxford University Press, 139–165.

Schadla-Hall, Tim. 1999. 'Editorial: Public Archaeology'. *European Journal of Archaeology* 2(2): 147–158.

Schadla-Hall, Tim. 2004. Appropriate Archaeology. In *Public Archaeology*, ed. Nick Merriman. London: Routledge, 255–271.

Scott, Karen. 2012. *Measuring Wellbeing: Toward Sustainability*. London: Routledge.

Sedlacek, Tomas. 2011. *Economics of Good and Evil. The Quest for Economic Meaning from Gilgamesh to Wall Street*. Oxford: Oxford University Press.

Seligman, Martin. 2011. *Flourish*. New York: Simon and Schuster.

Sen, Amartya. 1980. 'Equality of What?'. In *Tanner Lectures on Human Values*, ed. Sterling McMurrin. Cambridge: Cambridge University Press.

Sen, Amartya. 1993. 'Capability and Wellbeing'. In *The Quality of Life*, ed. Martha Nussbaum and Amartya Sen. New York: Oxford University Press, 30–53.

Shanks, Michael, and Christopher Tilley. 1987. *Social Theory and Archaeology*. Oxford: Polity Press.

Sharma, Ritu. 2017. *Global Youth Wellbeing Index*. International Youth Foundation.

Shennan, Stephen. 1988. 'Introduction'. In *Archaeology Approaches to Cultural Identity*, ed. Stephen Shennan. London: Unwin Hyman, 1–32.

Simpson, Faye. 2010. *The Values of Community Archaeology: A Comparative Assessment between the UK and US*. Oxford: Oxbow, British Archaeological Review.

Simpson, Faye, and Howard Williams. 2008. 'Evaluating Community Archaeology in the UK'. *Public Archaeology* 7(2): 69–90.

Smiraglia, Christina. 2015. 'Museum Programming and Mood: Participant Responses to an Object-Based Reminiscence Outreach Program in Retirement Communities'. *Arts & Health* (Special Issue: Culture, Museums and Wellbeing), 7(3): 187–201.

Smith, Claire, Vincent Copley, Josephine Kylie Lower, Ania Kotaba, and Gary Jackson. 2022. 'Using Archaeology to Strengthen Indigenous Social, Economic and Economic Wellbeing'. In *Archaeology, Heritage and Wellbeing: Authentic, Powerful and Therapeutic Engagement with the Past*, ed. Paul Everill and Karen Burnell. Abingdon: Routledge, 119–144.

Smith, Laura Jane. 2004. *Archaeological Theory and the Politics of Cultural Heritage*. London: Routledge.

Smith, Laura Jane. 2006. *The Uses of Heritage*. London: Routledge.

Smith, Laura Jane. 2021. *Emotional Heritage: Visitor Engagement at Museums and Heritage Sites*. London: Routledge.

Smith, Laura Jane, and Emma Waterton. 2009. *Communities and Archaeology*. London: Duckworth.

Smith, Laura Jane, and Emma Waterton. 2012. 'Constrained by Commonsense: The Authorized Heritage Discourse in Contemporary Debates'. In *The Oxford Handbook of Public Archaeology*, ed. Robin Skeates, Carol McDavid, and John Carman. Oxford: Oxford University Press, 154–171.

Social Value UK. 2018. *What Is Social Value*. Retrieved from http://www.socialvalueuk.org/what-is-social-value/ [accessed 2 January 2020].

Sofaer, Joanna, Ben Davenport, Marie-Louise Stig Sorensen, Eirini Gallou, and David Uzzell. 2021. 'Heritage Sites, Wellbeing and Value'. *International Journal of Heritage and Society* 27(11): 1117–1132.

Solent University. 2022. *AMPORA*. Retrieved from https://www.solent.ac.uk/research-innovation-enterprise/research-at-solent/projects-and-awards/march-plus-project [accessed 9 December 2022].

Spitzer, Robert, Kurt Kroenke, and Janet Williams. 1999. 'Patient Health Questionnaire Primary Care Study: Validation and Utility of a Self-Reported Version of PRIME-MD—The PHQ Primary Patient Health Questionnaire Primary Care Study'. *Journal of American Medical Association* 282(18): 1737–1744.

Stiglitz, Joseph, Amartya Sen, and Jean-Paul Fitoussi. 2009. *Report by the Commission on the Measurement of Economic Performance and Social Progress*. Paris: Commission on the Measurement of Economic Performance and Social Progress.

Stone, Peter. 2004. 'Introduction: Education and the Historic Environment into the Twenty First Century'. In *Education and the Historic Environment*, ed. Don Henson, Peter Stone, and Mike Corbishley. London: Routledge, 1–12.

Strauss, Anselm, and Juliet Corbin. 1990. *Basics of Grounded Theory Methods*. Beverly Hills: Sage Publishers.

Tacon, Paul. 2019. 'Connecting to the Ancestors: Why Rock Art Is Important for Indigenous Australians and Their Well-Being'. *Rock Art Research* 36(1): 5–14.

Tacon, Paul, and Sarah Baker. 2019. 'New and Emerging Challenges to Heritage and Wellbeing: A Critical Review'. *Heritage* 2: 1300–1315.

Tennant, Ruth, Louise Hiller, Ruth Fishwick, Stephen Platt, Stephen Joseph, Scott Weich, Jane Parkinson, Jenny Secker, and Sarah Stewart-Brown. 2007. 'The Warwick-Edinburgh Mental Well-Being Scale (WEMWBS): Development and UK Validation'. *Health and Quality of Life Outcomes* 5(1): 63.

Thin, Neil. 2009. 'Why Anthropology Can Ill Afford to Ignore Well-Being'. In *Pursuits of Happiness: Well-Being in Anthropological Perspective*, ed. Gordon Mathews and Carolina Izquierdo. New York: Berghahn Books, 22–44.

Thomas, Julian. 2004. *Archaeology and Modernity*. London: Routledge.

Thomas, Richard. 2004. 'Archaeology and Authority in the Twenty-First Century'. In *Public Archaeology*, ed. Nick Merriman. London: Routledge, 191–202.

Thomson, Linda, Eric Ander, Usha Menon, Anne Lanceley, and Helen Chatterjee. 2011. 'Evaluating the Therapeutic Effects of Museum Object Handing with Hospital Patients: A Review of Initial Trial of Wellbeing Measures'. *Journal of Applied Arts and Health* 2(1): 37–56.

Thomson, Linda, Eric Ander, Usha Menon, Anne Lanceley, and Helen Chattarjee. 2012. 'Quantitative Evidence for Wellbeing Benefits from a Heritage-in-Health Intervention with Hospital Patients'. *International Journal of Art Therapy* 17(2): 63–79.

Thomson, Linda, and Helen Chatterjee. 2013. *UCL Museums Wellbeing Measures Toolkit*. London: UCL. Retrieved from https://www.ucl.ac.uk/biosciences/culture-nature-health-research/ucl-creative-wellbeing-measures [accessed 23 February 2023].

Thomson, Linda, and Helen Chatterjee. 2014. 'Assessing Wellbeing Outcomes for Arts and Heritage Activities: Development of a Museum Wellbeing Measures Toolkit'. *Journal of Applied Arts & Health* 5(1): 29–50.

Thomson, Linda, and Helen Chatterjee. 2015. 'Measuring the Impact of Museum Activities on Wellbeing: Developing the Museum Wellbeing Measures Toolkit'. *Journal of Museum Management and Curatorship* 30(1): 44–62.

Tilley, Christopher. 1994. *A phenomenology of Landscape*. Oxford: Berg.

Topp, Christian, Soren Østergaard, Susan, Søndergaard, and Per Bech. 2015. 'The WHO-5 Well-Being Index: A Systematic Review of the Literature'*Psychotherapy and Psychosomatics*. 84: 167–176.

Tovar-Restrepo, Marcela. 2012. *Castoriadis, Foucault, and Autonomy: New Approaches to Subjectivity, Society and Social Change*. London: Bloomsbury.

Trigger, Bruce. 1989. *The History of Archaeological Thought*. Cambridge: Cambridge University Press.

Ucko, Peter. 1989. *Academic Freedom and Apartheid*. London: Duckworth.

UNSECO. 1995. *World Commission on Culture and Development, Our Creative Diversity*. Paris: UNESCO.

UNESCO. 2014. *World Heritage Convention Tentative Lists: Monuments of Srirangapatna Island Town*. Retrieved from https://whc.unesco.org/en/tentativelists/5895/ [accessed 10 August 2022].

UNESCO. 2016. *Strategy on Education for Health and Well-being: Contributing to the Sustainable Development Goals*. Paris: UNESCO.

UK Government Housing and Communities. 2022. *Levelling Up in the United Kingdom*. Retrieved from https://assets.publishing.service.gov.uk/government/uploads/system/uploads/attachment_data/file/1095544/Executive_Summary.pdf [accessed 18 March 2023].

United Nations. 2011. *General Assembly. A/RES/65/309: Happiness: Towards a Holistic Approach to Development*. Retrieved from https://documents-dds-ny.un.org/doc/UNDOC/GEN/N11/420/70/PDF/N1142070.pdf?OpenElement [accessed 9 December 2019].

United Nations. 2015. *Transforming Our World: The 2030 Sustainable Development Agenda*. New York: United Nations. Retrieved from https://sustainabledevelopment.un.org/post2015/transformingourworld/publication [accessed 23 February 2023].

United States Census Bureau. 2020. *2020 Census*. Retrieved from https://www.census.gov/programs-surveys/decennial-census/decade/2020/2020-census-main.html [accessed 10 August 2022].

United States Department of Health and Human Services. 2000. *Healthy People 2010*. Washington, DC: US Government Printing Office.

Ura, Karma, Sabina Alkire, Tshoki Zangmo, and Karma Wangdi. 2012. *An Extension Analysis of GNH Index*. Thimphu: Centre for Bhutan Studies.

Uzzell, David. 1989. 'The Hot Interpretation of War and Conflict'. In *Heritage Interpretation*, Vol. 1: *The Natural and Built Environment*, ed. D. Uzzell. London: Belhaven Press. 33–47.

Wallace, Jennifer. 2004. *Digging the Dirt: The Archaeological Imagination*. London: Gerald Duckworth & Co. Ltd.

Walshe, Diarmaid, Richard Osgood, and Martin Brown. 2012. 'Archaeology as Rehabilitation'. *British Archaeology* 122: 38–43.

Ward, George. 2015. *CEP Discussion Paper No 1343 April 2015 Is Happiness a Predictor of Election Results?* Centre for Economic Performance. Retrieved from http://cep.lse.ac.uk/pubs/download/dp1343.pdf [accessed 6 November 2017].

Waterloo Uncovered. 2019. Waterloo Uncovered Impact Report 2019. Retrieved from https://waterloouncovered.com/wp-content/uploads/2020/06/waterloo-uncovered-impact-report-2019.pdf [accessed 2 January 2023].

Waterson, Merlin. 1994. *The National Trust: The First Hundred Years*. London: BBC Books and National Trust (Enterprises) Limited.

Watson, David, Lee Clark, and Auke Tellegen. 1988. 'Development and Validation of Brief Measures of Positive and Negative Affect: The PANAS Scale'. *Journal of Personality and Social Psychology* 54(6): 1063–1070.

Welsh Government. 2022. Winter of Wellbeing. Retrieved from www.gov.wales/winter-wellbeing-sporting-activities-get-young-people-active-winter [accessed 20 March 2022].

Wheatley, Daniel, and Craig Bickerton. 2017. 'Subjective Well-Being and Engagement in Arts, Culture and Sport'. *Journal of Culture and Economics* 41(1): 23–45.

Whelan, Gayle. 2015. 'Understanding the Social Value and well-Being Benefits Created by Museums: A Case for Social Return on Investment Methodology'. *Arts & Health* 7(3): 216–230.

Whitaker, Patrick. 1995. *Managing to Learn*. London: Cassell.

Willey, Gordon, and Philip Phillips. 1958. *Method and Theory in American Archaeology*. Chicago: University of Chicago Press.

Williams, Allison. 2007. 'Introduction: The Continuing Maturation of the Therapeutic Landscape Concept'. In *Therapeutic Landscapes*, ed. Allison Williams. London: Ashgate, 1–12.

Wodak, Ruth. 2012. *Critical Discourse Analysis*. London: Sage Publications.

Woollard, Vicky. 2004. 'Caring for the Visitor'. In *Running a Museum: A Practical Handbook*, ed. Patrick Boylan. Paris: International Council of Museums.

World Bank Group. 2023. *Population India*. Retrieved from https://data.worldbank.org/indicator/SP.POP.TOTL.MA.ZS?end=2021&locations=IN&start=1960&view=chart [accessed 20 March 2023].

World Health Organization. 1946. *Constitution*. Geneva: WHO.

World Health Organization. 1998. *Wellbeing Measures in Primary Health Care/The Depcare Project*. Copenhagen: WHO Regional Office for Europe.

World Health Organization. 2004. *Promoting Mental Health: Concepts, Emerging Evidence, Practice* (Summary Report). Geneva: World Health Organization.

World Health Organization. 2022. The Geneva Charter. Retrieved from https://cdn.who.int/media/docs/default-source/health-promotion/geneva-charter-4-march-2022.pdf?sfvrsn=f55dec7_21&download=true [accessed 23 February 2023].

World Health Organization and Commission on the Social Determinants of Health. 2008. *Closing the Gap in a Generation: Health Equality through Action of the Social Determinants of Health*. Retrieved from http://whqlibdoc.who.int/publications/2008/9789241563703_eng.pdf [accessed 23 February 2023].

Yap, Mandy, and Yu Eunice 2016. 'Operationalising the Capability Approach: Developing Culturally Relevant Indicators of Indigenous Wellbeing—An Australian Example'. *Oxford Development Studies* 44(3): 315–331.

Zbranca, Rarita, Mafalda Dâmaso, Oana Blaga, Korelia Kiss, Denisa Dascl, Dana Yakobson, and Oana Pop. 2022. *CultureForHealth Report. Culture's Contribution to Health and Well-Being. A Report on Evidence and Policy Recommendations for Europe*. CultureForHealth. Culture Action Europe. Retrieved from https://www.cultureforhealth.eu/app/uploads/2023/02/Final_C4H_FullReport_small.pdf [accessed 9 March 2023].

Index

Since the index has been created to work across multiple formats, indexed terms for which a page range is given (e.g., 52–53, 66–70, etc.) may occasionally appear only on some, but not all of the pages within the range.

Figures are indicated by an italic "*f*".